# ROCKETS

DR. ROBERT H. GODDARD

# ROCKETS

## Robert H. Goddard

COMPRISING
*"A Method of Reaching Extreme Altitudes"*
AND
*"Liquid-Propellant Rocket Development"*

(Two Articles Bound as One)

DOVER PUBLICATIONS, INC.
Mineola, New York

Published in the United Kingdom by David & Charles, Brunel House, Forde Close, Newton Abbot, Devon TQ12 4PU.

### Bibliographical Note

This Dover edition, first published in 2002, is an unabridged republication of the work originally published in 1946 by the American Rocket Society, New York, which consists of two articles bound as one. They are "A Method of Reaching Extreme Altitudes" and "Liquid-Propellant Rocket Development," originally published in 1919 and 1936, respectively, by the Smithsonian Institution, Washington, D.C.

### Library of Congress Cataloging-in-Publication Data

Goddard, Robert Hutchings, 1882–1945
    Rockets / Robert H. Goddard.
        p. cm.
    "Compromising 'A method of reaching extreme altitudes' and 'Liquid-propellant rocket development'."
    Originally published: New York : American Rocket Society, 1946.
    ISBN 0-486-42537-1 (pbk.)
    1. Rocketry. 2. Rockets (Aeronautics). 3. Atmosphere, Upper—Rocket observations. 4. Liquid propellant rockets. I. Goddard, Robert Hutchings, 1882–1945. Method of reaching extreme altitudes. II. Goddard, Robert Hutchings, 1882–1945. Liquid-propellant rocket development. III. Title.

TL782 .G62 2002
621.43'56—dc21

2002071623

Manufactured in the United States of America
Dover Publications, Inc., 31 East 2nd Street, Mineola, N.Y. 11501

# PREFACE

The two most significant publications in the history of rockets and jet propulsion are "A Method of Reaching Extreme Altitudes" and "Liquid-Propellant Rocket Development."

Both of these reports, published originally by the Smithsonian Institution—the first in 1919 and the second in 1936—have been unavailable for some time. They are here re-published by the American Rocket Society, by arrangement with the estate of Dr. Robert H. Goddard and the Smithsonian Institution, with a new foreword by Dr. Goddard which was written a little over four months before his death on August 10, 1945.

Perhaps it is unnecessary to point out the significance of these two early technical reports, issued at a time when rockets and jet propulsion, now so universally accepted as an established field of engineering, were little known and less respected. Though rockets have been in use for more than seven hundred years, it was not until Goddard undertook his lifelong work on the development of high altitude rockets, starting about 1909, that really modern scientific and engineering methods began to be applied in this field.

It is a tribute to the fundamental nature of Dr. Goddard's work that these reports, though now several years old, are filled with data of vital importance to all jet propulsion and rocket engineers. These reports form one of the most important technical contributions of our time.

Dr. Goddard may truly be called the father of modern rocketry. Before 1920 he had developed most of the basic concepts in rocket and jet propulsion engineering. He had already obtained patents, many of them fundamental in character, and had called world-wide attention to the possibilities of jet propulsion both as applied to projectiles and aircraft.

Because of the importance of the Goddard papers—the only formal reports of his work so far published—it has been a matter of concern to engineers that they have been so long out of print. The American Rocket Society is happy to be able to provide them now to all who have need of the data developed by Dr. Goddard's work.

# CONTENTS

# FOREWORD

*to the new edition of "A Method of
Reaching Extreme Altitudes" and
"Liquid-Propellant Rocket Development"*

The writer's two papers on rockets, published by the Smithsonian Institution in 1919 and 1936, are so different in content that it seems best to write a separate foreword for each.

## A METHOD OF REACHING EXTREME ALTITUDES

It is interesting to compare the "state of the art" at the time the above paper was written with the viewpoints as well as the accomplishments of the present moment. Smokeless powder rockets, during World War II, have grown in size and controllability, but have not given much higher jet velocities than were earlier obtained. Multiple charge, or reloading, powder rockets appear still to be in the status mentioned on page 68.

Liquid fuel rockets, which were presented through a patent reference with little emphasis, have progressed very considerably, and have caused the assumed weight ratio of 1/15 for empty to loaded rocket to appear much more attainable. The jet velocities for liquids, however, notwithstanding the higher energies, are still in the smokeless powder range.

So far as general principles and conclusions go, both appear to be as sound as ever. Thus high jet velocity is still of as much importance for long ranges as light rocket construction. Moreover, the jet velocity should be high for the entire propulsion period, even though a low jet velocity in the early part of the ascent would give a greater energy efficiency. This low jet velocity would obviously be disadvantageous in requiring a larger initial fuel load.

The development of the theory on the basis of the conservation of momentum remains to the writer preferable to using energy. Especially worth while is the procedure of varying the thrust in such a way as to maintain an optimum velocity at each point of the flight, in order to minimize the fuel load. The multiple, or step, rocket principle would be as good as ever in maintaining a low ratio of rocket to fuel, if the application of this method had been found necessary.

As was suggested, a start from a great elevation would reduce considerably the initial mass of propellant required. A further reduction, not mentioned in the early paper, could be made by using some of the tremendous energy of the blast to give the rocket a substantial initial velocity.

Finally, the subject of projection from the earth, and especially a mention of the moon, must still be avoided in dignified scientific and engineering circles, even though projection over long distances on the earth's surface no longer calls for quite so high an elevation of eyebrows. In this connection, it would appear best merely to repeat the concluding paragraph on page 57, which holds good as much now as then, and to remember that, from some points of view, twenty-five years is not so very long, after all.

## LIQUID-PROPELLANT ROCKET DEVELOPMENT

The 1919 paper indicated the theoretical possibility of obtaining great ranges by means of well designed rockets; showed also that fairly high jet velocities were obtainable, and mentioned progress in the construction of a solid cartridge, magazine-type rocket, in which the weight of explosive charge to rocket could be large.

The next step was to develop such a rocket that would operate for a considerable time, and have satisfactory flight characteristics. While this work was in progress, some simple combustion tests were carried out with liquid oxygen and a number of liquid fuels. These tests indicated that liquid oxygen could be handled without much difficulty, and that satisfactory combustion could probably be obtained in a chamber, or motor, of comparatively simple design.

Proving stand tests were accordingly carried out with liquid oxygen and gasoline from 1920 to 1926, when a liquid fuel rocket flight was obtained. Those who took part experienced a lift in spirit for the reason that the rocket functioned, but the distance attained did not seem great enough to warrant calling attention to the event.

Further short flights were made up to 1929, when Colonel Charles A. Lindbergh came to Clark University to learn of the status of the work. He concluded that the liquid fuel rocket flights, even though short, had significance, and encouraged Mr. Daniel Guggenheim to make a grant for a further development. The work begun in this way was continued later under the Daniel and Florence Guggenheim Foundation, the details being as set forth in the 1936 paper.

Flights of a thousand feet or so, with a relatively heavily loaded rocket, showed at once that automatic stability in flight would be needed before making any attempt to increase the range. Gyroscopically controlled vanes in the rocket blast gave, first, an indication of flight correction and finally continued correcting from side to side, at approximately a 10° displacement angle, as long as the chamber was in operation. Beyond this point the trajectory was curved, as it had previously been during propulsion without gyro control. The flame was clear in some tests, and accompanied by smoke in others, the latter showing the correcting best in flight photographs.

It was believed by 1935 that the results indicated the possibility of making a rocket ascent which would be high compared with the records for both meteorology and aeronautics, when a comprehensive report would, of course, be in order. Nevertheless, it seemed worth while to publish the 1936 paper as a progress report, in order to show what had already been accomplished.

It was thereafter found necessary to carry further the work on flight characteristics in order to obtain correction during the coasting period, as otherwise the speed when the parachute was released was likely to be excessive. This work led to flight correction to within 3°, during both the powered and coasting periods, together with reliable parachute release at the peak of the flight.

There remained the production of a very light weight rocket, namely one having a pump-turbine drive and light tanks, using the guiding system that had already been worked out with pressure tank rockets. Preliminary flights with such a rocket were under way in 1941, at the time the development was dropped because of other, more urgent war problems.

ROBERT H. GODDARD

May 1, 1945

A METHOD OF REACHING EXTREME ALTITUDES was finished in manuscript by Dr. Goddard on May 26, 1919. It consisted primarily of a paper written in 1916 at Clark College, amplified by additional data and calculations based on subsequent observation and research. The 1916 paper was prepared by the physicist as a means of enlisting the support of the Smithsonian Institution for his work. He later told associates that very little changing was necessary to prepare it for publication; his original theories and calculations had proved astonishingly accurate. Though dated 1919, the publication was released by the Smithsonian early in January, 1920, when copies were made available to the press. Only 1,750 copies of this paper were printed, but they were sufficient to launch a new era, creating almost instant world-wide interest, and setting off trains of research and speculation which are undoubtedly to have profound effects on the future history of the world.

*Reproduced here by facsimile printing exactly as originally published by the Smithsonian Institution in 1919*

SMITHSONIAN MISCELLANEOUS COLLECTIONS
VOLUME 71, NUMBER 2

# A METHOD OF REACHING EXTREME ALTITUDES

(With 10 Plates)

BY

ROBERT H. GODDARD
Clark College, Worcester, Mass.

(Publication 2540)

CITY OF WASHINGTON
PUBLISHED BY THE SMITHSONIAN INSTITUTION
1919

# PREFACE

The theoretical work herein presented was developed while the writer was at Princeton University in 1912-13, the basis of the calculations being the assumption that, if nitrocellulose smokeless powder were employed as propellant in a rocket, under such conditions as are here explained, an efficiency of 50 per cent might be expected.

Actual experimental investigations were not undertaken until 1915-16, at which time the tests concerning ordinary rockets, steel chambers and nozzles, and trials *in vacuo*, were performed at Clark University. The original calculations were then repeated, using the data from these experiments, and both the theoretical and experimental results were submitted, in manuscript, to the Smithsonian Institution, in December, 1916. This manuscript is here presented in the original form, save for the notes at the end which are now added.

A grant of $5,000 from the Hodgkins Fund, Smithsonian Institution, under which work is being done at present, was advanced toward the development of a reloading, or multiple-charge rocket, herein explained in principle, and this work was begun at the Worcester Polytechnic Institute in 1917, and was later undertaken as a war proposition. It was continued, from June, 1918, up to very nearly the time of signing of the armistice, at the Mt. Wilson Observatory of the Carnegie Institution of Washington, where most of the experimental results were obtained.

In connection with the present publication, I take pleasure in thanking Dr. A. G. Webster for the facilities of the shop and laboratory at Clark University, used in the preliminary experiments herein described. I also take this opportunity of expressing my gratitude to the Smithsonian Institution, for its support and encouragement in the later work.

ROBERT H. GODDARD.

CLARK COLLEGE,
  WORCESTER, MASSACHUSETTS,
    May 26, 1919.

# A METHOD OF REACHING EXTREME ALTITUDES

## By ROBERT H. GODDARD

### (With 10 Plates)

### OUTLINE

A search for methods of raising recording apparatus beyond the range for sounding balloons (about 20 miles) led the writer to develop a theory of rocket action, in general (pp. 6 to 11), taking into account air resistance and gravity. The problem was to determine the minimum initial mass of an ideal rocket necessary, in order that on continuous loss of mass, a final mass of one pound would remain, at any desired altitude.

An approximate method was found necessary, in solving this problem (pp. 10 to 11), in order to avoid an unsolved problem in the Calculus of Variations. The solution that was obtained revealed the fact that surprisingly small initial masses would be necessary (table VII, p. 46) *provided the gases were ejected from the rocket at a high velocity,* and also provided that *most of the rocket consisted of propellant material.* The reason for this is, very briefly, that the velocity enters *exponentially* in the expression for the initial mass. Thus if the velocity of the ejected gases be increased five fold, for example, the initial mass necessary to reach a given height will be *reduced to the fifth root* of that required for the lesser velocity. (A simple calculation, p. 50, shows at once the effectiveness of a rocket apparatus of high efficiency.)

It was obviously desirable to perform certain experiments: First, with the object of finding just how inefficient an ordinary rocket is, and secondly, to determine to what extent the efficiency could be increased in a rocket of new design. The term " efficiency " here means the ratio of the kinetic energy of the expelled gases to the heat energy of the powder, the kinetic energy being calculated from the average velocity of ejection, which was obtained indirectly by observations on the *recoil* of the rocket.

It was found that not only does the powder in an ordinary rocket constitute but a small fraction of the total mass ($\frac{1}{4}$ or $\frac{1}{3}$), but that, furthermore, the efficiency is only 2 per cent, the average velocity of ejection being about 1,000 ft./sec. (table I, p. 12). This was true even in the case of the Coston ship rocket, which was found to have a range of a quarter of a mile.

Experiments were next performed with the object of increasing the average velocity of ejection of the gases. Charges of dense smokeless powder were fired in strong steel chambers (fig. 2, p. 13), these chambers being provided with smooth tapered nozzles, the object of which was to obtain the work of expansion of the gases, much as is done in the De Laval steam turbine. The efficiencies and velocities obtained in this way were remarkably high (table II, p. 15), the highest efficiency, or rather " duty," being over 64 per cent, and the highest average velocity of ejection being slightly under 8,000 ft./sec., which exceeds any velocity hitherto attained by matter in appreciable amounts.

These velocities were proved to be real velocities, and not merely effects due to reaction against the air, by firing the same steel chambers *in vacuo*, and observing the recoil. The velocities obtained in this way were not much different from those obtained in air (table III, p. 30).

It will be evident that a heavy steel chamber, such as was used in the above-mentioned experiments, could not compete with the ordinary rocket, even with the high velocities which were obtained. If, however, *successive charges* were fired in the *same chamber*, much as in a rapid fire gun, *most of the mass of the rocket could consist of propellant*, and the superiority over the ordinary rocket could be increased enormously. Such reloading mechanisms, together with what is termed a " primary and secondary " rocket principle, are the subject of certain United States Letters Patent (p. 6). Inasmuch as these two features are self-evidently operative, it was not considered necessary to perform experiments concerning them, in order to be certain of the practicability of the general method.

Regarding the heights that could be reached by the above method; an application of the theory to cases which the experiments show must be realizable in practice indicates that a mass of one pound could be elevated to altitudes of 35, 72, and 232 miles; by employing initial masses of from 3.6 to 12.6, from 5.1 to 24.3, and from 9.8 to 89.6 pounds, respectively (table VII, p. 46). If a device of the Coston ship rocket type were used instead, the initial masses would be of the order of magnitude of *those above, raised to the 27th power.* It should be understood that if the mass of the recording instruments alone were one pound, the entire final mass would be 3 or 4 pounds.

Regarding the possibility of recovering apparatus upon its return, calculations (pp. 51 and 53) show that the times of ascent and descent will be short, and that a small parachute should be sufficient to ensure safe landing.

Calculations indicate, further (pp. 54 to 57), that with a rocket of high efficiency, consisting chiefly of propellant material, it should be possible to send small masses even to such great distances as to escape the earth's attraction.

In conclusion, it is believed that not only has a new and valuable method of reaching high altitudes been shown to be *operative in theory*, but that the experiments herein described *settle all the points upon which there could be reasonable doubt*.

The following discussion is divided into three parts:

Part   I.   Theory.
Part   II.   Experiments.
Part   III.   Calculations, based upon the theory and the experimental results.

## IMPORTANCE OF THE SUBJECT

The greatest altitude at which soundings of the atmosphere have been made by balloons, namely, about 20 miles, is but a small fraction of the height to which the atmosphere is supposed to extend. In fact, the most interesting, and in some ways the most important, part of the atmosphere lies in this unexplored region; a means of exploring which has, up to the present, not seriously been suggested.

A few of the more important matters to be investigated in this region are the following: the density, chemical constitution, and temperature of the atmosphere, as well as the height to which it extends. Other problems are the nature of the aurora, and (with apparatus held by gyroscopes in a fixed direction in space) the nature of the $\alpha$, $\beta$, and $\gamma$ radioactive rays from matter in the sun as well as the ultra-violet spectrum of this body.

Speculations have been made as to the nature of the upper atmosphere—those by Wegener [1] being, perhaps, the most plausible. By estimating the temperature and percentage composition of the gases present in the atmosphere, Wegener calculates the partial pressures of the constituent gases, and concludes that there are four rather distinct regions or spheres of the atmosphere in which certain gases predominate: the troposphere, in which are the clouds; the stratosphere, predominatingly nitrogen; the hydrogen sphere; and the

---

[1] A. Wegener. Phys. Zeitschr. 12, pp. 170-178; 214-222, 1911.

geocoronium sphere. This highest sphere appears to consist essentially of an element, " geocoronium," a gas undiscovered at the surface of the earth, having a spectrum which is the single aurora line, $557\mu\mu$, and being 0.4 as heavy as hydrogen. The existence of such a gas is in agreement with Nicholson's theory of the atom, and its investigation would, of course, be a matter of considerable importance to astronomy and physics as well as to meteorology. It is of interest to note that the greatest altitude attained by sounding balloons extends but one-third through the second region, or stratosphere.

No instruments for obtaining data at these high altitudes are herein discussed, but it will be at once evident that their construction is a problem of small difficulty compared with the attainment of the desired altitudes.

# A METHOD OF REACHING EXTREME ALTITUDES

### By ROBERT H. GODDARD

## PART I.  THEORY

### METHOD TO BE EMPLOYED

It is possible to obtain a suggestion as to the method that must be employed from the fundamental principles of mechanics, together with a consideration of the conditions of the problem.  We are at once limited to an apparatus which reacts against matter, this matter being carried by the apparatus in question.  For the entire system we must have: First, action in accordance with Newton's Third Law of Motion; and, secondly, energy supplied from some source or sources must be used to give kinetic and potential energy to the apparatus that is being raised; kinetic energy to the matter which, by reaction, produces the desired motion of the apparatus; and also sufficient energy to overcome air resistance.

We are at once limited, since sub-atomic energy is not available, to a means of propulsion in which jets of gas are employed.  This will be evident from the following consideration: First; the matter which, by its being ejected furnishes the necessary reaction, must be taken with the apparatus in reasonably small amounts.  Secondly, energy must be taken with the apparatus in as large amounts as possible.  Now, inasmuch as the maximum amount of energy associated with the minimum amount of matter occurs with chemical energy, both the matter and the energy for reaction must be supplied by a substance which, on burning or exploding, liberates a large amount of energy, and permits the ejection of the products that are formed.  An ideal substance is evidently smokeless powder, which furnishes a large amount of energy, but does not explode with such violence as to be uncontrollable.

The apparatus must obviously be constructed on the principle of the rocket.  An ordinary rocket, however, of reasonably small bulk, can rise to but a very limited altitude.  This is due to the fact that the part of the rocket that furnishes the energy is but a rather small fraction of the total mass of the rocket; and also to the fact that only a part of this energy is converted into kinetic energy of the mass which is expelled.  It will be expected, then, that the ordinary rocket is an inefficient heat engine.  Experiments will be described below which show that this is true to a surprising degree.

By the application of several new principles, an efficiency manyfold greater than that of the ordinary rocket is possible; experimental demonstrations of which will also be described below. Inasmuch as these principles are of some value for military purposes, the writer has protected himself, as well as aerological science in America, by certain United States Letters Patent; of which the following have already been issued:

> 1,102,653
> 1,103,503
> 1,191,299
> 1,194,496.

The principles concerning efficiency are essentially three in number. The first concerns thermodynamic efficiency, and is the use of a smooth nozzle, of proper length and taper, through which the gaseous products of combustion are discharged. By this means the work of expansion of the gases is obtained as kinetic energy, and also complete combustion is ensured.

The second principle is embodied in a reloading device, whereby a large mass of explosive material is used, a little at a time, in a small, strong, combustion chamber. This enables high chamber pressures to be employed, impossible in an ordinary paper rocket, and also permits most of the mass of the rocket to consist of propellant material.

The third principle consists in the employment of a primary and secondary rocket apparatus, the secondary (a copy in minature of the primary) being fired when the primary has reached the upper limit of its flight. This is most clearly shown, in principle, in United States Patent No. 1, 102,653.

By this means the large ratio of propellant material to total mass is kept sensibly the same during the entire flight. This last principle is obviously serviceable only when the most extreme altitudes are to be reached. In order to avoid damage when the discarded casings reach the ground, each should be fitted with a parachute device, as explained in United States Patent No. 1,191,299.

Experiments will be described below which show that, by application of the above principles, it is possible to convert the rocket from a very inefficient heat engine into the most efficient heat engine that ever has been devised.

## STATEMENT OF THE PROBLEM

Before describing the experiments that have been performed, it will be well to deduce the theory of rocket action in general, in order to show the tremendous importance of efficiency in the attainment of very high altitudes. A statement of the problem will therefore be made, which will lead to the differential equation of the motion. An approximate solution of this equation will be made for the initial mass required to raise a mass of one pound to any desired altitude, when said initial mass is a minimum.

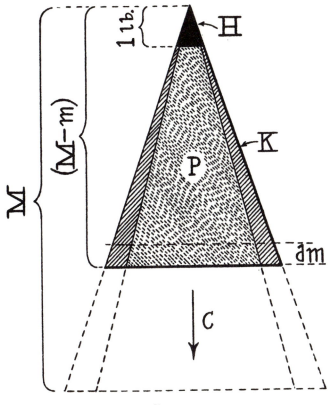

FIG. I.

A particular form of ideal rocket is chosen for the discussion as being very amenable to theoretical treatment, and at the same time embodying all of the essential points of the practical apparatus. Referring to figure 1, a mass H, weighing one pound is to be raised as high as possible in a vertical direction [1] by a rocket formed of a cone, P, of propellant material, surrounded by a casing K. The

---

[1] See note 10 at end of paper.

material P is expelled downward with a constant velocity, c. It is further supposed that the casing, K, drops away continuously as the propellant material P burns, so that the base of the rocket always remains plane. It will be seen that this approximates to the case of a rocket in which the casing and firing chamber of a primary rocket are discarded after the magazine has been exhausted of cartridges, as well as to the case in which cartridge shells are ejected as fast as the cartridges are fired.

Let us call

$M =$ the initial mass of the rocket,

$m =$ the mass that has been ejected up to the time, t,

$v =$ the velocity of the rocket, at time t,

$c =$ the velocity of ejection of the mass expelled,

$R =$ the force, in absolute units, due to air resistance,

$g =$ the acceleration of gravity,

$dm =$ the mass expelled during the time dt,

$k =$ the constant fraction of the mass dm that consists of casing K, expelled with zero velocity relative to the remainder of the rocket, and

$dv =$ the increment of velocity given the remaining mass of the rocket.

The differential equation for this ideal rocket will be the analytical statement of Newton's Third Law, obtained by equating the momentum at a time t to that at the time $t + dt$, plus the impulse of the forces of air resistance and gravity,

$$(M - m)v = dm(1 - k)(v - c) + vkdm$$
$$+ (M - m - dm)(v + dv) + [R + g(M - m)]dt.$$

If we neglect terms of the second order, this equation reduces to

$$c(1 - k)dm = (M - m)dv + [R + g(M - m)]dt. \qquad (1)$$

A check upon the correctness of this equation may be had from the analytical expression for the Conservation of Energy, obtained by equating the heat energy evolved by the burning of the mass of propellant, $dm(1 - k)$, to the additional kinetic energy of the system produced by this mass plus the work done against gravity and air resistance during the time dt. The equation thus derived is found to be identical with equation (1).

### REDUCTION OF EQUATION TO THE SIMPLEST FORM

In the most general case, it will be found that R and g are most simply expressed when in terms of v and s. In particular, the quantity R, the air resistance of the rocket at time t, depends not only upon the density of the air and the velocity of the rocket, but also upon the cross section, S, at the time t. The cross section, S, should obviously be as small as possible; and this condition will be satisfied *at all times,* provided it is the following function of the mass of the rocket (M−m),

$$S = A(M-m)^{\frac{2}{3}} \qquad (2)$$

where A is a constant of proportionality. This condition is evidently satisfied by the ideal rocket, figure 1. Equation (2) expresses the fact that the shape of the rocket apparatus is at all times similar to the shape at the start; or, expressed differently, S must vary as the square of the linear dimensions, whereas the mass (M−m) varies as the cube. Provision that this condition may approximately be fulfilled is contained in the principle of primary and secondary rockets.

The resistance, R, may be taken as independent of the length of the rocket by neglecting " skin friction." For velocities exceeding that of sound this is entirely permissible, provided the cross section is greatest at the head of the apparatus, as shown in United States Patent No. 1,102,653.

The quantities R, g, and v, are evidently expressible most simply in terms of the altitude s, provided the cross-section S is also so expressed, giving, in place of equation (1)

$$c(1-k)dm = (M-m)dv + \frac{1}{v(s)}[R(s)+g(s)(M-m)]ds. \qquad (3)$$

### RIGOROUS SOLUTION FOR MINIMUM M AT PRESENT IMPOSSIBLE

The success of the method depends entirely upon the possibility of using an initial mass, M, of explosive material that is not impracticably large. It amounts to the same thing, of course, if we say that the mass ejected up to the time t (*i. e.,* m) must be a minimum, conditions for the existence of a minimum being involved in the integration of the equation of motion.

That a minimum mass, m, exists when a required mass is to be given an assigned upward velocity at a given altitude is evident intuitively from the following consideration : If, at any intermediate altitude, the velocity of ascent be very great, the air resistance R (depending upon the square of the velocity) will also be great. On the other hand, if the velocity of ascent be very small, force will be required to overcome gravity for a long period of time. In both cases the mass necessary to be expelled will be excessively large.

Evidently, then, the velocity of ascent must have some special value at each point of the ascent. In other words it is necessary to determine an unknown function f(s), defined by

$$v = f(s),$$

such that m is a minimum.

It is possible to put $f(s)$ and $\dfrac{df(s)}{ds}$ ds in place of v and dv, in equation (3), and to obtain m by integration. But in order that m shall be a minimum, δm must be put equal to zero, and the function $f(s)$ determined. The procedure necessary for this determination presents a new and unsolved problem in the Calculus of Variations.

## SOLUTION OF THE MINIMUM PROBLEM BY AN APPROXIMATE METHOD

In order to obtain a solution that will be sufficiently exact to show the possibilities of the method, and will at the same time avoid the difficulties involved in the employment of the rigorous method just described, use may be made of the fact that if we divide the altitude into a large number of parts, let us say, n, we may consider the quantities R, g, and also the acceleration, to be *constant over each interval*.

If we denote by a the constant acceleration defined by $v = at$ in any interval, we shall have, in place of the equation of motion (3), a linear equation of the first order in m and t, as follows:

$$\frac{dm}{dt} = \frac{(M-m)(a+g)+R}{c(1-k)}$$

the solution of which, on multiplying and dividing the right number by $(a+g)$, is

$$m = e^{-\frac{a+g}{c(1-k)}t} \cdot \frac{M(a+g)+R}{a+g}\left[\left[\int e^{\frac{a+g}{c(1-k)}t}\left(\frac{a+g}{c(1-k)}\right)dt + C\right]\right]$$

$$= e^{-\frac{a+g}{c(1-k)}t} \cdot \frac{M(a+g)+R}{a+g}\left[e^{\frac{a+g}{c(1-k)}t} + C\right],$$

where C is an arbitrary constant.

This constant is at once determined as $-1$ from the fact that m must equal zero when $t = 0$.

We then have

$$m = \left(M + \frac{R}{a+g}\right)\left[1 - e^{-\frac{a+g}{c(1-k)}t}\right]. \tag{5}$$

This equation applies, of course, to each interval, R, g, and a. being considered constant. We may make a further simplification if, for each interval, we *determine what initial mass, M, would be required when the final mass in the interval is one pound.* The initial mass at the beginning of the first interval, or what may be called the " total initial mass," required to propel the apparatus through the n intervals will then be the *product of the n quantities* obtained in this way.

If we thus place the final mass $(M-m)$, in any interval equal to unity, we have $M=m+1$ and when this relation is used in equation (5), we have for the mass at the beginning of the interval in question

$$M = \frac{R}{a+g}\left(e^{\frac{a+g}{c(1-k)}t} - 1\right) + e^{\frac{a+g}{c(1-k)}t} \qquad (6)$$

Now the initial mass that would be required to give the one pound mass the same velocity at the end of the interval, if R and g had both been *zero,* is, from (6)

$$M = e^{\frac{at}{c(1-k)}}. \qquad (7)$$

The ratio of equation (6) to equation (7) is a measure of the additional mass that is required for overcoming the two resistances, R and g; and when this ratio is least, we know that M is a minimum for the interval in question. The " total initial mass" required to raise one pound to any desired altitude may thus be had as the product of the minimum M's for each interval, obtained in this way.

From equations (6) and (7) we see at once the importance of high efficiency, if the " total initial mass" is to be reduced to a minimum. Consider the exponent of e. The quantities a, g and t depend upon the particular ascent that is to be made, whereas $c(1-k)$ depends entirely upon the efficiency of the rocket, c being the velocity of expulsion of the gases, and k, the fraction of the entire mass that consists of loading and firing mechanism, and of magazine. In order to see the importance of making $c(1-k)$ as large as possible, suppose that it were decreased tenfold. Then $e^{\frac{a+g}{c(1-k)}t}$ would be *raised to the 10th power,* in other words, the mass for each interval would be the *original value multiplied by itself ten times.*

## PART II.  EXPERIMENTS

### EFFICIENCY OF ORDINARY ROCKET

The average velocity of ejection of the gases expelled from two sizes of ordinary rocket were determined by a ballistic pendulum. The smaller rockets, C, plate 1, figure 1, averaged 120 grams, with a powder charge of 23 grams; and the larger, S, the well-known Coston ship rocket, weighed 640 grams, with a powder charge of 130 grams. Plate 1, figure 2, shows the rockets as compared with a yard-stick, Y.

The ballistic pendulum, plate 2, figures 1 and 2, was a massive compound pendulum weighing 70.64 Kg. (155 lbs.) with a half period of 4.4 seconds; large compared with the duration of discharge of the rockets. The efficiencies were obtained from the average velocity of ejection of the gases, found by the usual ballistic pendulum method, together with the heat value of the powder of the rockets, obtained by a bomb calorimeter for the writer by a Worcester chemist.

The results of these experiments are given in the following table:

TABLE I

| Type of rocket | Efficiency | Mean efficiency | Velocity corresponding to mean efficiency |
|---|---|---|---|
| Common............ | 2.54% | | |
| "          ............ | 1.45 | | |
| "          ............ | 1.49 | | |
| "          ............ | 1.95 | 1.86% | 957.6 ft./sec. |
| Coston ship........ | 1.75% | | |
| "          ........ | 2.27 | | |
| "          ........ | ·2.62 | 2.21% | 1029.25 ft./sec. |

It will be seen from the above table that the efficiency of the ordinary rocket is close to 2 per cent"; slightly less for the smaller, and slightly more for the larger, rockets; and also that the average velocity of the ejected gases is of the order of 1,000 ft./sec. It was found by experiment that a Coston ship rocket, lightened to 510 grams by the removal of the red fire, had a range of a quarter of a mile, the highest point of the trajectory being slightly under 490 feet. A range as large as this is rather remarkable in view of the surprisingly small efficiency of this rocket.

## EXPERIMENTS IN AIR WITH SMALL STEEL CHAMBERS

An apparatus was next constructed, with a view to increasing the efficiency, embodying three radical changes, namely, the use of smokeless powder, of much higher heat value than the black powder employed in ordinary rockets; the use of a strong steel chamber, to permit employment of high pressures; and the use of a tapered nozzle, similar to a steam turbine nozzle, to make available the work of expansion.

Two sizes of chamber were used, one $\frac{1}{2}$ inch diameter, and one 1 inch diameter. The inside and outside diameters of the smaller chamber, figure 2 (a), were, respectively, 1.28 cm. and 3.63 cm. The

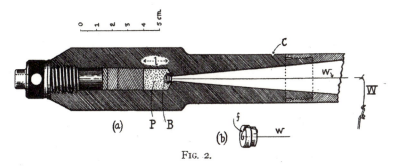

Fig. 2.

nozzle, polished until very smooth, was of 8 degrees taper, and was adapted to permit the use of two extensions of different lengths. The length of the chamber, as the distance 1 in the figure will be called, could be altered by putting in or removing cylindrical tempered steel plugs of various lengths, held in place by the breech block. Plate 2, figure 3, is a photograph of the chamber, with the nozzle of medium length in position. Plate 2, figure 4, shows the same, taken apart; and figure 5 shows the chamber with the medium nozzle replaced by the long nozzle.

Two small chambers were used, practically identical in all respects; one of the soft tool steel, and one of best selected nickel steel gun barrel stock, treated to give 100,000 lbs. tensil strength, for which the writer wishes to express his indebtedness to the Winchester Repeating Arms Company.

The charge of powder, P, figure 2, was fired electrically, by a hot wire in the following way: A fine copper wire, w, 0.12 mm. diameter, passed through the wadding, figure 2 (b), consisting of two disks of stiff cardboard, and this copper wire joined a short length of platinum or platenoid wire of 0.1 mm. diameter, f, extending across the inner part of the wadding, in contact with the powder. To the other end of this platinum wire, a short length of the copper wire passed to the side of the wadding, and made electrical contact with the wall of the chamber. A fine steel wire, W, 0.24 mm. in diameter, served to pull the copper wire, w, tightly enough to prevent contact of the latter with the nozzle. The wire, W, was so held that, although it exerted a pull on the wire w, it nevertheless offered no resistance in the direction of motion of the ejected gases.

Two dense smokeless powders were used: Du Pont pistol powder No. 3, a very rapid dense nitrocellulose powder, and "Infallible" shotgun powder, of the Hercules Powder Company. The heat values in all cases were found by bomb calorimeter.[1] All determinations were made in an atmosphere of carbon dioxide, in order to avoid any heat due to the oxygen of the air. The average heat values were the following:

Powder, in ordinary rocket.......... 545.0 calories/gm.
Powder, in Coston ship rocket....... 528.3
Du Pont Pistol No. 3.............. 972.5
"Infallible" ...................... 1238.5

The ballistic pendulum used in determining the average velocity of ejection, for the small chambers, consisted essentially of a plank, B, plate 3, figure 1, carrying weights, and supporting the chamber, or gun, C, in a horizontal position. This plank was supported by fine steel wires in such a manner that it remained horizontal during motion. In order to make certain that the plank actually was horizontal in all positions, a test was frequently made by mounting a small vertical mirror on the plank, with its plane perpendicular to the axis of the gun, and observing the image of a horizontal object—as a lead pencil—held several feet away while the pendulum was swinging. Current for firing the charge was lead through two drops of mercury to wires on the plank. A record of the displacements was made by a stilus consisting of a steel rod, S, pointed and hardened at the lower end. This rod slid freely in a vertical brass sleeve, attached to the under side of the plank, and made a mark upon a smoked glass strip, G. In this way the first backward and forward displacements of the pendulum were recorded, and the elimination of friction was thereby made possible.

The data and results of these experiments are given in table II, in which d is the displacement corrected for friction.

---

[1] It was found necessary to use a sample exceeding a certain mass, as otherwise the heat value depended upon the mass of the sample.

TABLE II

SMALL CHAMBER

| Experiment No. | Chamber | Length of Chamber ℓ | Total Mass M | Length of Nozzle | Kind of Powder | Mass of Powder | Mass of Wadding and Wire | $d_1$ | $d_2$ | corrected $d$ | Length of Pendulum Wire | Velocity Km./sec. | Velocity Ft./sec. | Efficiency E |
|---|---|---|---|---|---|---|---|---|---|---|---|---|---|---|
| 1 | Soft Steel | 0.69 cm | 3540.1 gm | Medium | Du Pont | 0.7795 gm | .0345 gm | 11.55 cm | 11.41 cm | 11.62 cm | 79.15 cm | 1.781 | 5843 | 39.01% |
| 2 | " | 0.69 | 3541.9 | " | " | 0.7060 | .0385 | 10.30 | 10.19 | 10.35 | 79.15 | 1.738 | 5703 | 37.16 |
| 3 | " | 1.01 | 3538.8 | " | " | 1.0025 | .0370 | 15.80 | 15.70 | 15.85 | 79.15 | 1.907 | 6257 | 44.73 |
| 4 | " | 0.69 | 3541.9 | " | Infallible | 0.8247 | .0395 | 13.60 | 13.50 | 13.65 | 79.15 | 1.976 | 6484 | 37.15 |
| 5 | " | 1.01 | 3538.8 | " | " | 1.2015 | .0380 | 20.55 | 20.46 | 20.59 | 79.50 | 2.082 | 6832 | 41.88 |
| 6 | " | 0.69 | 3547.9 | Short | Du Pont | 0.7074 | .0370 | 9.43 | 9.38 | 9.45 | 79.50 | 1.585 | 5203 | 30.93 |
| 7 | " | 0.69 | 3540.1 | " | Infallible | 0.8533 | .0370 | 12.59 | 12.53 | 12.62 | 79.50 | 1.766 | 5793 | 30.12 |
| 8 | " | 0.69 | 3540.1 | " | Du Pont | 0.6825 | .0355 | 9.35 | 9.31 | 9.37 | 79.50 | 1.626 | 5336 | 32.54 |
| 9 | " | 1.01 | 3645.8 | Long | Infallible | 1.2397 | .0370 | 20.18 | 20.10 | 20.22 | 79.50 | 2.045 | 6709 | 40.39 |
| 10 | " | 1.01 | 3645.8 | " | Du Pont | 0.9625 | .0365 | 14.20 | 14.10 | 14.25 | 79.50 | 1.834 | 6018 | 41.38 |
| 11 | " | 0.69 | 3648.93 | " | " | 0.7361 | .0386 | 10.22 | 10.10 | 10.28 | 79.50 | 1.704 | 5592 | 35.74 |
| 12 | " | 0.69 | 3535.9 | Medium | Infallible | 0.8985 | .0391 | 13.90 | 13.83 | 13.94 | 79.50 | 1.850 | 6069 | 33.05 |
| 13 | " | 0.69 | 3645.8 | Long | " | 0.9068 | .0396 | 13.85 | 13.80 | 13.87 | 79.50 | 1.882 | 6177 | 34.24 |
| 14 | " | 0.69 | 3533.9 | Medium | Du Pont | 0.7465 | .0373 | 10.07 | 10.00 | 10.10 | 79.50 | 1.609 | 5279 | 31.38 |
| 29 | Ni Steel | 0.69 | 3553.5 | " | Infallible | 1.0264 | .0445 | 17.95 | 17.85 | 18.00 | 79.50 | 1.969 | 6460 | 37.44 |
| 44 | " | 1.01 | 6273.5 | " | " | 1.2731 | .0420 | 12.58 | 12.38 | 12.68 | 79.50 | 2.127 | 6981 | 43.73 |
| 46 | " | 0.69 | 6270.5 | " | " | 1.4849 | .0402 | 14.78 | 14.68 | 14.93 | 79.50 | 2.154 | 7064 | 44.78 |

LARGE CHAMBER

| Experiment No. | Chamber | Length of Chamber ℓ | Total Mass M | Length of Nozzle | Kind of Powder | Mass of Powder | Mass of Wadding and Wire | $d_1$ | $d_2$ | corrected $d$ | Length of Pendulum Wire | Velocity Km./sec. | Velocity Ft./sec. | Efficiency E |
|---|---|---|---|---|---|---|---|---|---|---|---|---|---|---|
| 51 | Cr – Ni Steel | 2.28 cm | 19324.0 gm | 16.29 cm | Du Pont | 8.0522 gm | .3184 gm | | | 5.02 cm | | 2.290 | 7515 | 64.53% |
| 52 | " | 2.28 | 19324.0 | 16.29 | Infallible | 9.0259 | .3271 | | | 7.08 | | 2.434 | 7987 | 57.25 |

The velocities and efficiencies were obtained from the usual expression for the velocity in which a ballistic pendulum, with the bob constantly horizontal, is used, namely,

$$v = \frac{M}{m} \sqrt{2gl(1 - \cos\theta)},$$

where

        $M$ = the total weight of the bob,
        $m$ = the mass ejected; powder plus wadding,
        $l$ = the length of the pendulum,
        $\theta$ = the angle through which the pendulum swings,
        $g$ = the acceleration of gravity.

The cosine of $\theta$ was corrected for friction by observing the two first displacements $d_1$ and $d_2$ and obtaining therefrom

$$d = d_1 \sqrt{\frac{d_1}{d_2}}.$$

It will be noticed that the highest velocity was obtained with " Infallible " powder, and was over 7,000 ft./sec. The corresponding efficiency was close to 50 per cent. In view of the fact that this velocity, corresponding to c in the exponents of equations (6) and (7), is sevenfold greater than for an ordinary rocket, it is easily seen that the employment of a chamber and nozzle such as has just been described must make an enormous reduction in initial mass as compared with that necessary for an ordinary rocket.

As a matter of possible interest, photographs were taken at night of the flash which accompanied the explosions produced by firing the small chamber. These are given on plates 4 and 5. Plate 3, figure 2, shows the set-up for these experiments; the camera being in the same position as when the flashes were photographed. The white marks, above the flash, are strips of cardboard, nailed to a long stick at intervals of 10 cm. and constituting a comparison scale, one end of which was directly above the " muzzle " of the gun. This scale was illuminated, before the charge was fired, by a small electric flash lamp held in front of each strip for a moment; which lamp also illuminated a card bearing the number of the experiment.

The photographs bring out a curious fact; *i. e,* that the " flash " appears in most instances to be at a considerable distance in front of the nozzle. This is easily understood if we admit that the velocity of the ejected gases is very high just as the gases pass out of the nozzle, but becomes very quickly reduced nearly to zero by the air. In other words we may consider that the gases pass from the nozzle in an extremely short time—far too short to affect the photographic plate; and that it is only when the velocity has been considerably reduced that the " flash " is photographed.

In experiment 11, a suggestion of this high-velocity portion of the flash is seen, which, it will be noticed, is less in diameter than the end of the nozzle. It should be remarked that it was only by accident that the nozzle was illuminated by the flashes in experiments 9 and 11 in such a way as to be seen in the photograph.

An interesting phenomenon connected with firing the chamber in air is the loudness of the sound produced by a comparatively small amount of powder. This is, however, to be expected, inasmuch as the energy is entirely spent in producing motion of the air, whereas in the ordinary rifle, a large fraction of the energy of the powder becomes kinetic energy of the bullet.

### EXPERIMENTS WITH LARGE CHAMBER

Inasmuch as all the steel chambers employed in the preceding experiments were of the same internal diameter (1.26 cm.), it was considered desirable that at least a few experiments should be performed with a larger chamber, first, in order to be certain that a large chamber is operative; and secondly, to see if such a chamber is not even more efficient than a small chamber. This latter is to be expected for the reason that heat and frictional losses should increase as the *square* of the linear dimensions of the chamber; and hence increase in a less proportion than the mass of powder that can be used with safety, which will vary as the *cube* of the linear dimensions. Evidence in support of this expectation has already been given. Thus, for ordinary rockets, the larger rocket has the higher efficiency, as evident from table I.

The large chamber was of nickel-alloy steel (Samson No. 3A), of 115,000 lbs. tensil strength, for which the writer takes opportunity of thanking the Carpenter Steel Company. This chamber had inside diameter, and diameter of throat, both twice as large as those of the chambers previously used; but the thickness of wall of the chamber and the taper of the nozzle were, however, the same. The inside of the nozzle was well polished. Figure 3 shows a section of the chamber; the outer boundary being indicated by dotted lines, P being the powder, and W the wadding. It will be noticed that the wadding is just twice the size of that previously used.

The mounting of the chamber, for the experiments, is shown in plate 6, figure 1. The chamber was held in the lower end of a

3½ foot length of 2-inch pipe, P, by set-screws. Within this pipe, above the chamber, was fastened a length of 2-inch steel shafting, to increase the mass of the movable system. This system was supported by a half-inch steel pin, E.

On firing, the recoil lifted the above system vertically upward against gravity, the extent of this lift, or displacement, being recorded by a thin lead pencil, slidable in a brass sleeve set in the pipe at right angles to the pin E. The point of the pencil was pressed against a vertical cardboard, C, by the action of a small spring. This method of measuring the impulse of the expelled gases will be called the " direct-lift " method; and the theory is given in Appendix A, page 60.

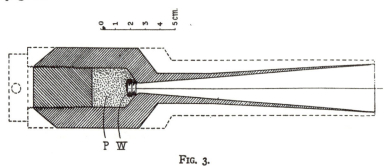

Fig. 3.

Although rebound of the gases from the ground would probably have been negligible, such rebound was eliminated by a short plank, D, covered with a piece of heavy sheet iron, and supported at an angle of 45° with the horizontal. This served to deflect the gases to one side.

The results of two experiments, 51 and 52, with this large chamber, are given in table II. In experiment 51, with Du Pont powder, the powder was packed rather loosely. Any increase in internal diameter was inappreciable, certainly under 0.01 mm. In experiment 52, the Infallible powder was somewhat compressed. After firing, the chamber was found to be slightly bulged for a short distance around the middle of the powder chamber, the inside diameter being increased from 2.6 cm. to 2.7 cm., and the outside diameter from 5.08 cm. to 5.14 cm. The efficiency (64.53 per cent) in experiment 51, and the velocity (7,987 ft./sec.) in experiment 52 were, respectively, the highest obtained in any of the experiments.

The conclusions to be drawn from these two experiments are: First, that large chambers can be operated, under proper conditions, without involving undue pressures; and secondly, that large chambers, even with comparatively short nozzles, are more efficient and give higher velocities than small chambers.

It is obvious that large grains of powder should be used in large chambers if dangerous pressures are to be avoided. The bulging in experiment 52 is to be explained by the grains of powder being too small for a chamber of the size under consideration. It is possible, however, that pressures even as great as that developed in experiment 52 could be employed in practice provided the chamber were of "built-up" construction. A similar result might possibly be had if several shots had been fired, of successively increasing amounts of powder. The result of this would have been a hardening of the wall of the chamber by stretching. Such a phenomenon was observed with the soft steel chamber already described, which was distended by the first few shots of Infallible powder, but thereafter remained unchanged with loads as great as those first used.[12]

## EXPERIMENTS IN VACUO

### INTRODUCTORY

Having obtained average velocities of ejection up to nearly 8,000 ft./sec. in air, it remained to determine to what extent these represented reaction against the air in the nozzle, or immediately beyond. Although it might be supposed that the reaction due to the air is small, from the fact that the air in the nozzle and immediately beyond is of small mass, it is by no means self-evident that the reaction is zero. For example, when dynamite, lying on an iron plate, is exploded, the particles which constituted the dynamite are moved very rapidly upward, and the reaction to this motion bends the iron plate downward; but reaction of the said particles against the air as they move upward may also play an important rôle in bending the iron. The experiments now to be described were undertaken with the view of finding to what extent, if any, the "velocity in air" was a fictitious velocity. The experiments were performed with the smaller soft tool steel and nickel-steel chambers that have already been described.

### METHOD OF SUPPORTING THE CHAMBER IN VACUO

For the sake of convenience, the chamber, or gun, should evidently be mounted in a vertical position, so that the expelled gases are shot downward, and the chamber is moved upward by the reaction, either being lifted bodily, or suspended by a spring and set in vibration.

The whole suspended system was therefore designed to be contained in a 3-inch steel pipe, all the essential parts being fastened to a cap, fitting on the top of this pipe. This was done not only for the sake of convenience in handling the heavy chamber, but also from the fact that the only joint that would have to be made air-tight for each shot would be at the 3-inch cap.

The means of supporting the chamber from the cap is shown in plate 6, figure 2, and plate 7, figure 1, the apparatus being shown dismantled in plate 7, figure 2. Two $\frac{3}{8}$-inch steel rods, R, R, were threaded tightly by taper (pipe) threads into the cap, C. These rods were joined by a yoke, at their lower ends, which served to keep them always parallel. Two collars, or holders, H and H', free to slide along the rods R, R, held the chamber or gun, by three screws in each holder. The inner ends of the screws of the lower holder were made conical, and these fitted into conical depressions, c, figure 2(a), drilled in the side of the gun, so that the lower holder could thus be rigidly attached to the gun. This was made necessary in order that lead sleeves, fitting the gun and resting upon the lower holder H', could be used to increase the mass of the suspended system. Three such sleeves were used, the two largest being moulded around thin steel tubes which closely fitted the gun. The rods R, R, were lubricated with vaseline. Two $\frac{1}{8}$-inch steel pins were driven through the rods R, R, just above the yoke Y, in order that the latter could not be driven off by the fall of the heavy chamber and weights when direct-lift was employed.

In the experiments in which the chamber and lead sleeves were suspended by a spring, the latter was hooked at its upper end to a screw-eye fixed in the cap C. The lower end of the spring was hooked through a small cylinder of fiber. A record of the displacements of the suspended system was made by a stilus, S, plate 6, figure 2, in the upper holder H. This stilus was kept pressed against a long narrow strip of smoked glass, G, by a spring of fine steel wire. This strip of smoked glass was held between two clamps, fastened to a rod, the upper end of which was secured to the cap C, and the lower end to the yoke, Y. Except for the largest charges used, it was possible to measure the displacements on both sides of the zero position, and thereby to calculate the decrement and eliminate friction.

When the chamber was suspended by a spring, a deflection as large as a centimeter was unavoidably produced merely by placing the cap C on the 3-inch pipe or removing it, although, in all cases the system would return to within one millimeter (usually much less than this) of the zero position after being displaced. In order to avoid any such displacement as that just mentioned, an eccentric clamp K, plate 7, figure 1, was employed to keep the suspended system rigidly in its zero position during assembling and dismounting the apparatus.

This clamp consisted of an eccentric rod, K, free to turn in a hole in the cap C, the lower end being held in a bearing in the yoke Y. Through the upper end of this rod was pinned a small rod K', at right angles to K. The surface of the rod K was smeared with a mixture of bee's wax, resin, and Venice turpentine; and the hole in the cap through which K projected was rendered air-tight by wax of the same composition.

The suspended system was assembled while the cap C was held by a support touching its under side. When the assembling was complete, the wax was heated by a small alcohol blow torch until it was soft, then a rubber band was slipped around the rod K' and the outlet pipe E. A trial showed that the cap could now be put in place on the pipe and removed, without the suspended system moving appreciably. After the cap C was in position on the pipe, the rubber band was removed, and the wax heated until the rod K could be turned out of engagement with the holders H, H'. After a shot had been fired, the clamp was again placed in operation until the system had been taken from the 3-inch pipe and the smoked glass removed.

The circuit which carried the electric current to ignite the charge consisted of the insulated wire, W, plate 7, figure 1, which passed through a tapered plug of shellacked hard fiber, in the cap, C, thence through a glass tube to the yoke Y, to which it was fastened. Below the yoke it was wrapped with insulating tape, except at the lower end where it was shaped to hold the 0.24 mm. steel wire, attached to the fine copper wire from the wadding. From the chamber the current passed up the rods R, R and out of the cap, around which was wrapped a heavy bare copper wire, V, which together with W, constituted the terminals of the circuit. It should be mentioned, in passing, that a small amount of black powder, B, figure 2(a), placed over the platinum fuse-wire on the wadding, was found necessary as a primer in order to ignite dense smokeless powders *in vacuo*.

In order to make the joint, between the cap and the pipe, air-tight during a determination, the following device was adopted. The outside of the cap, C, and also a lock nut, were both turned down to the same diameter. The lock nut was made fast to the pipe. These were then painted on the outside with melted wax consisting of equal parts bee's wax and resin with a little Venice turpentine.

When a determination was to be made, the cap was screwed into position, a wide rubber band was slipped over the junction between cap and lock nut, and the outside of this rubber band was heated with an alcohol blast torch. The result was a joint, for all practical purposes, absolutely air tight, which could, nevertheless, be dismounted at once after pulling off the rubber band.

### THEORY OF THE EXPERIMENTS IN VACUO

The expressions for the velocity of the expelled gases are easily obtained for the two types of motion of the suspended system that were employed, namely, simple harmonic motion produced by a spring, and direct lift.

*Simple harmonic motion.*—Results obtained with simple harmonic motion (slightly damped, of course) were naturally more accurate than with direct lift, as it was impossible in the latter case to eliminate friction. The theory, for simple harmonic motion, in which account is taken of friction is described in Appendix B, page 60. The spring was one made to specifications, particularly as regards the magnitude of the force per-cm.-increase-in-length by the Morgan Spring Company of Worcester, Mass. Care was taken to make certain that in no experiment was the extension of the spring reduced to such a low value as not to lie upon the rectilinear line part of the calibration curve.

*Direct lift.*—The theory of the motion, in this case, has already been given under Appendix A, page 60. In this case it might be assumed that a correction could be made for friction by multiplying the displacement s, by some particular decrement, $\sqrt{\dfrac{d_1}{d_2}}$, obtained in the experiments with simple harmonic motion, that might reasonably apply. This, as will be shown below, was found to give results in good agreement for the two types of motion, if the direct lift was about 2 cm.; but not if it was much larger. It was found that very little frictional resistance was experienced when the mass M was raised by hand, provided the axis of the gun were kept strictly vertical, but a very considerable resistance was experienced if the axis was inclined to one side so that the holders H, H' rubbed against the rods R, R. This sidewise pressure did not take place when the spring was used. It was also found that the trace upon the smoked glass was always slightly sinuous, with direct lift, and straight with the spring. The simple harmonic motion was, therefore, much the more preferable, but could not be used when the powder charges were large.

### MEANS OF ELIMINATING GASEOUS REBOUND

It should be remembered that the real object of the vacuum experiments is to ascertain what the reaction experienced by the chamber would be, if a given charge of powder were fired in the chamber many miles above the earth's surface. A container is therefore necessary, which, for the purpose at hand, approaches most nearly a container of unlimited capacity. A length of 3-inch pipe, closed at the ends, is evidently unsuitable, because the gas, fired from one end, is sure to rebound from the other end with considerable velocity, and hence to produce a much larger displacement than ought really to be observed. Moreover, any tank of finite size must necessarily produce a finite amount of rebound, from the fact that the whole action is equivalent to liberating suddenly, in the tank, one or two liters of gas at atmospheric pressure.

There are two possible methods for reducing the velocity of the gas sufficiently to produce a negligible rebound: a *disintegration method,* whereby the stream is broken up into many small streams, sent in all directions (*i. e.,* virtually reconverted into heat) ; and secondly, a *friction method,* whereby the individual stream remains moving in one direction, but is gradually slowed down by friction against a solid surface.

As will be shown below, accurate results were obtained by the first method, in what may be called the " cylindrical " tank ; and these results were checked satisfactorily by the second method, in what will be called the " circular " tank.

The cylindrical tank was 10 feet 5 inches high and weighed about 500 lbs. It consisted of a 6-foot length T, figure 4, and plate 8, figure 1, of 12-inch steel pipe, with threaded caps on the ends. Entering the upper cap at a slight angle was the 3-inch pipe P, $4\frac{1}{2}$ feet long which supported the cap C of plate 6, figure 2, and plate 7, figure 1. The 12-inch pipe was sawn across at the dotted line $T_0$, so that any device could be placed in the interior of this tank, or removed from it, as desired. The upper section of the tank was lifted off as occasion demanded by a block and tackle. The two ends to be joined were first painted with the wax previously described ; and after the tank had been assembled, the joint was painted on the outside with the same wax, W, and the entire tank thereafter painted with asphalt varnish.

This tank was used under three conditions:

1st. Tank empty, with the elbow E to direct the gas into a swirl such that the gas, while in motion, would not tend to return up the pipe P. In this case, some rebound was to be expected from this elbow. This expectation was realized in practice.

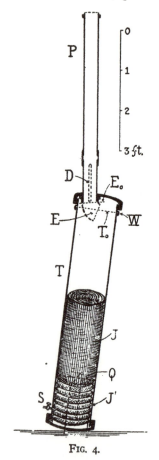

FIG. 4.

2d. Tank empty, and elbow cut off along the dotted line $E_0$. In this case, more rebound was to be expected than in Case 1, which was borne out in practice.

3d. Elbow E cut at $E_0$, and tank half filled with $\frac{1}{2}$-inch square-mesh wire fencing. Two separate devices constructed of this wire fencing were used one above the other. The gas first passed through an Archimedes spiral, J, of 2-foot fencing, comprising eight turns.

held apart by iron wires bound into the fencing. This construction allowed most of the gas to penetrate the spiral to a considerable distance before being disturbed, and, of course, eliminated regular reflection. The second device, J', placed under the first, consisted of a number of 12-inch circular disks of the same fencing, bound to two quarter-inch iron rods, Q, by iron wires. These disks were spaced one inch apart. The three upper disks were single disks, the next lower two were double, with the strands extending in different directions, the next two were triple, and the lowest disk of all, two inches from the bottom of the tank, was composed of six individual disks. This lower device necessarily offered large resistance to the passage of the gas; yet strong rebound from any part of it was prevented by the spiral just described. With this third arrangement, small rebound was to be expected, which also was borne out in practice.

This tank was exhausted by way of a stopcock at its lower end, S; and air was also admitted through this same stopcock.

The circular tank, plate 8, figure 2, was 10 feet high and weighed about 200 lbs. It consisted of a straight length of 3-inch pipe, carefully fitted, and welded autogenously, to a four-foot, 3-inch, U-pipe. The straight pipe entered the U-pipe on the inner side of the latter, and at as sharp an angle as possible. Another similar U-pipe, was bolted to the first by flanges, with $\frac{1}{16}$-inch sheet rubber packing between.

In this tank, the gases were shot down the straight pipe, entered the upper U-pipe at a small angle, thus avoiding any considerable rebound, and thence passed around the circular part—not returning up the straight pipe until the velocity had been greatly reduced by friction.

In order to make the time, during which the velocity was being reduced, as long as possible, the pipes were carefully cleaned of scale. They were first pickled, and then cleaned by drawing through them, a number of times: first, a scraper of sheet iron; second, a stiff cylindrical bristle brush, and finally a cloth. All but the most firmly adhering scale was thereby removed. Further, care was taken to cut the hole in the rubber washers, between the flanges, so wide that compression by the flanges would not spread the rubber into the pipe, and thereby obstruct the flow of gas.

Notwithstanding all these precautions, evidence was had that the gases became stopped very rapidly. This was to be expected inasmuch as there is solid matter, namely, the wadding and wire, that is ejected with the gas, which accumulates with each successive shot. This solid matter must offer considerable frictional resistance to motion along the U-pipe, and, since the mass of gas is only of the order of a gram, must necessarily act to stop the flow in a very short time. This interval of time was great enough, however, so that this second method afforded a satisfactory check upon the first method.

A possible modification of the above two methods would have been to provide some sort of trap-door arrangement whereby the gases, after having been reduced in speed in a container as just described, would have been prevented from returning upward into the 3-inch pipe P by this trap, which would be sprung at the instant of firing. In this way gaseous rebound would be entirely eliminated. It was found, however, that results with the two methods already described could be checked sufficiently to make this modification unnecessary.

The tanks were exhausted by a rotary oil pump, No. 1, of the American Rotary Pump Company, supported by a water jet pump. In this way the pressure in the cylindrical tank could be reduced to 1.5 mm. of mercury in 25 minutes and to the same pressure in the circular tank, in 10 minutes. The pressures employed in the experiments ranged from 7.5 mm. to 0.5 mm.

### METHODS OF DETECTING AND MEASURING GASEOUS REBOUND

With the two tanks used in the experiments, it was obviously impossible to eliminate gaseous rebound entirely, from the fact that, even if the velocity of the gases is reduced to zero, there still remains the effect of introducing suddenly a certain quantity of gas into the tank. It became necessary, then, to devise some means of detecting, and, if possible, of measuring, the extent of the rebound.

Three devices were employed, one for detecting a *force* of rebound, and two for measuring the magnitude of the *impulse per unit area* produced by the rebounding gas. These latter devices, from the fact that quantitative measurements were possible with them, will be called " impulse-meters."

### TISSUE PAPER DETECTOR

The detector for indicating the force of the rebound consisted of a strip of delicate tissue-paper, I, plate 6, figure 2, and text figure 5 (a), 0.02 mm. thick, with its ends glued to an iron wire, W, as shown in figure 5 (a). This iron wire was fastened to the yoke Y, plate 7, figure 1, and held the tissue paper, with its plane horizontal, between the chamber and the wall of the 3-inch pipe, P. In many of the experiments, the paper was cut $\frac{1}{3}$ the way across in two places before being used, as shown by the dotted lines h in figure 5(a). Since the tissue paper has very little mass, the tearing depends simply upon the magnitude of the force that is momentarily applied, and not upon the force times its duration—*i. e.,* the impulse of the force. The

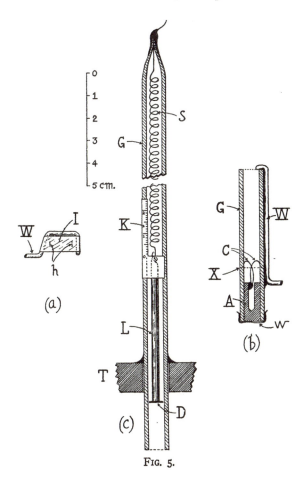

Fig. 5.

tissue paper will tear, then, if the force produced by the first upward rush of gas, past the chamber into the space in the 3-inch pipe above the chamber, exceeds a certain value. This first upward rush of gas will, of course, produce a greater force than any subsequent rush, as the gas is continually losing velocity. Even though the magnitude of the force that will just tear the tissue paper be not known, it may

safely be assumed that if the first upward rush does not tear the
paper, the force due to rebound that acts upon the gun must be small
compared with the impulse produced by the explosion of the powder.

It should be noted that the tissue paper tells nothing as to whether
or not there are a number of successive reflections or rebounds gradually decreasing in magnitude; neither does it give information concerning the *downward* pressure the gases exert upon the chamber
tending to decrease the displacement, after they have accumulated in
the space between the top of the chamber and the cap, C, plate 6,
figure 2.

### DIRECT-LIFT IMPULSE-METER

A section of the direct-lift impulse-meter is shown in figure 5(b).
It is also shown in the photograph plate 6, figure 2, at A. A small
cylinder A of aluminium of 1.46 grams mass, hollowed at one end
for lightness, was turned down to slide easily in a glass tube G. This
tube, G, was fastened by de Khotinsky cement to an iron wire W,
which was in turn fastened to the yoke Y, plate 7, figure 1, so that the
glass tube, G, was held in a vertical position, between the chamber
and the wall of the 3-inch pipe—similarly to the tissue paper. Two
small wires C, C, of spring brass were cemented to the top of the
aluminium cylinder, the free ends just touching on opposite sides of
the glass tube. The inside of the glass tube was smoked with cam-
phor smoke above the point marked X, so that a record was made of
any upward displacement of the aluminium cylinder. The cylinder
was prevented from dropping out of the glass tube by a fine steel
wire, w, cemented to the tube and extending across the lower end.

The theory of the direct-lift impulse-meter is given in Appendix C,
page 61. From the theory, we may derive an expression for the
ratio, Q, of the momentum given the gun by the gaseous rebound, to
the observed momentum of the suspended system.

There are two disadvantages of this form of impulse-meter. First,
friction acts unavoidably to reduce the displacement. Secondly, any
jar to which the apparatus is subjected on firing will cause the
aluminium cylinder to jump, and thus give a spurious displacement.
This latter fact rendered the meter useless for experiments in which
direct lift of the chamber took place, as there was always much jar
when the heavy chamber fell back, after being displaced upward.

This impulse-meter, it will be observed, gave a mean measurement
of any successive up-and-down rushes of gas.

## SPRING IMPULSE-METER

A section of the spring impulse-meter is shown in figure 5(c). The apparatus consisted of an aluminium disk, D, cemented to a lead rod, L, of combined mass 5.295 gms. supported by a fine brass spiral spring, S. The disk, D, was of a size sufficient to slide easily in a glass tube, G. The upper end of the spring protruded through a small hole in the glass tube, and was fastened at this point by de Khotinsky cement, it thus being easy to make the top of the lead rod level with the zero of a paper scale, K, pasted to the outside of the glass tube. A piece of white paper placed behind the tube, G, made the motion of the lead rod L very clearly discernible.

This impulse-meter was placed in a hole in the upper cap of the 12-inch pipe of the cylindrical tank, at D, figure 4, and plate 8, figure 1, the same distance from the wall of the 12-inch pipe as the center of the 3-inch pipe. It projected one inch through the 12-inch cap which was practically the same as the distance the 3-inch pipe projected. The tube, G, was kept in position in the cap by being wrapped tightly with insulating tape, the joint being finally painted with the wax already described.

The theory of the spring impulse-meter is given in Appendix D, page 62, where Q is the ratio already defined in connection with the direct-lift impulse-meter. There are two reasons why the ratio, Q, obtained in the Appendix should be larger than the true percentage at the top of the 3-inch pipe. In the first place, friction in the 3-inch pipe will decrease the velocity of the rebounding gas; and, further, the disk, D, figure 5, is fairly tight-fitting in the glass tube, G, whereas there is a considerable space between the gun and the 3-inch pipe, through which the gas may pass and, accumulating above, exert a *downward* pressure on the top of the gun.

One important advantage of the spring impulse-meter over that employing direct lift is that the former has very little friction, so that the readings are very reliable. Another advantage is that the displacement of the former will include without any uncertainty the effect of any number of rebounds following one another in rapid succession—*i. e.*, the effect of multiple reflections of the gas, if such reflections are present.

## EXPLANATION OF TABLE III

In the vacuum experiments, the soft steel chamber was used for Du Pont powder, and the nickel steel chamber for Infallible powder.

The three nozzles called short, medium, and long, were respectively, 9.64, 15.88, and 22.08 cm. from the throat to the muzzle.

**TABLE III**

| Experiment No. | Type of Motion | Length of Chamber ℓ | Total Mass $M+\frac{1}{2}m_0$ | Length of Nozzle | Kind of Powder | Mass of Powder | Mass of Wadding and Wire | Mass of Black Powder | Displacement $d_1$ | Displacement $d_i$ | Displacement corrected $d$ | Tank | Pressure Before | Pressure After | Paper Detector | Rebound Impulse to Total Impulse Q | Velocity Km/sec | Velocity Ft/sec | Efficiency |
|---|---|---|---|---|---|---|---|---|---|---|---|---|---|---|---|---|---|---|---|
| 15 | S.H.M. | 0.69 cm | 3156.9 gm | Long | Du Pont | 0.6747 gm | .0538 gm | .007 gm | 4.58 cm | 4.22 cm | 4.97 cm | cylindrical | 5.0 mm | 5.5 mm | — | .000 % | 1.711 | 5614 | 36.01 |
| 16 | " | 0.69 | 3156.9 | " | " | 0.6761 | .0526 | .007 | 4.78 | 4.70 | 4.82 | " | 5.0 | 10.0 | torn | .756 | 1.729 | 5671 | 36.75 |
| 17 | " | 0.69 | 3156.9 | " | " | 0.6913 | .0508 | .007 | 4.68 | 4.52 | 4.70 | " | 4.5 | 9.0 | " | .000 | 1.671 | 5481 | 34.33 |
| 18 | " | 0.69 | 3156.9 | " | " | 0.6929 | .0536 | .007 | 4.85 | 4.55 | 5.01 | " | 7.0 | 11.2 | " | .000 | 1.774 | 5821 | 58.72 |
| 19 | " | 0.69 | 3158.9 | " | " | 0.6741 | .0529 | .007 | 4.66 | 4.37 | 4.81 | " | 5.5 | 10.5 | " | .000 | 1.728 | 5663 | 36.11 |
| 20 | " | 0.69 | 3156.9 | " | " | 0.7161 | .0516 | .007 | 5.00 | 4.77 | 5.12 | " | 7.5 | 13.0 | not torn | .000 | 1.683 | 5524 | 34.86 |
| 21 | " | 0.69 | 3156.9 | " | " | 0.6495 | .0536 | .007 | 4.73 | 4.45 | 4.81 | " | 6.5 | 10.5 | " | .000 | 1.780 | 5840 | 38.97 |
| 22 | " | 0.69 | 3156.9 | " | " | 0.6679 | .0568 | .007 | 4.63 | 4.34 | 4.78 | circular | 1.5 | 13.5 | " | .560 | 1.719 | 5642 | 36.37 |
| 23 | " | 0.69 | 3156.9 | " | " | 0.6681 | .0537 | .007 | 4.45 | 4.13 | 4.59 | cylindrical | 1.5 | 5.5 | " | .000 | 1.653 | 5423 | 33.61 |
| 24 | " | 0.69 | 3156.9 | " | " | 0.6693 | .0556 | .007 | 4.68 | 4.48 | 4.78 | circular | 1.5 | 22.0 | " | .000 | 1.719 | 5642 | 36.38 |
| 25 | " | 0.69 | 2768.1 | Medium | " | 0.6998 | .0504 | .007 | 4.97 | 4.31 | 5.33 | cylindrical | 1.5 | 14.5 | " | .000 | 1.767 | 5801 | 38.46 |
| 26 | " | 0.69 | 2768.1 | Short | " | 0.6715 | .0530 | .007 | 4.70 | 3.85 | 5.19 | circular | 1.5 | 5.0 | " | .000 | 1.749 | 5740 | 37.65 |
| 27 | " | 0.69 | 2353.8 | Medium | Infallible | 0.6686 | .0510 | .007 | 5.05 | (*20) | 5.17 | cylindrical | 1.5 | 13.0 | " | — | 1.614 | 5296 | 32.05 |
| 28 | " | 0.69 | 2353.8 | " | " | 0.6673 | .0510 | .007 | 5.10 | (*20) | 5.22 | cylindrical | 1.5 | 10.5 | " | — | 1.630 | 5347 | 32.67 |
| 30 | " | 0.95 | 3339.6 | Medium | " | 0.9186 | .0556 | .010 | 7.37 | (*20) | 7.91 | circular | 1.5 | 21.0 | " | — | 2.405 | 7893 | 55.90 |
| 31 | Lift | 0.95 | 2020.7 | " | " | 0.9210 | .0518 | .012 | 4.80 | (*25) | 4.94 | cylindrical | 1.5 | 7.5 | torn | — | 1.997 | 6550 | 39.40 |
| 32 | " | 0.95 | 2020.7 | " | " | 0.9210 | .0601 | .020 | 5.87 | (*20) | 5.90 | circular | 1.5 | 21.0 | " | — | 2.191 | 7189 | 46.38 |
| 33 | " | 0.95 | 2020.7 | " | " | 0.9210 | .0625 | .020 | 5.30 | (*25) | 5.69 | circular | 1.5 | 25.5 | " | — | 2.127 | 6980 | 43.71 |
| 34 | " | 0.95 | 2020.7 | " | " | 0.9210 | .0648 | .020 | 5.50 | (*25) | 5.90 | cylindrical | 4.5 | 8.0 | not torn | — | 2.162 | 7093 | 45.15 |
| 35 | S.H.M. | 0.95 | 3339.6 | " | " | 0.9210 | .0614 | .020 | 7.22 | (*20) | 7.39 | " | 2.5 | 6.0 | " | — | 2.336 | 7665 | 52.73 |
| 36 | " | 0.95 | 3339.6 | " | " | 0.9210 | .0639 | .020 | 7.18 | (*20) | 7.35 | " | 1.5 | 9.5 | torn | — | 2.319 | 7610 | 51.96 |
| 37 | Lift | 0.95 | 2020.7 | Long | " | 0.9210 | .0619 | .020 | 5.19 | (*25) | 5.57 | " | 1.5 | 7.0 | not torn | — | 2.106 | 6911 | 42.85 |
| 38 | " | 0.95 | 2135.7 | Medium | " | 0.9210 | .0672 | .020 | 4.83 | (*25) | 5.18 | " | 1.5 | 7.0 | " | — | 2.136 | 7010 | 44.09 |
| 39 | " | 0.95 | 2135.7 | Short | Du Pont | 0.9210 | .0608 | .020 | 5.07 | (*25) | 5.45 | " | 1.5 | 7.0 | " | — | 2.202 | 7227 | 46.86 |
| 40 | " | 0.69 | 2023.4 | Medium | " | 0.6715 | .0576 | .007 | 2.03 | (*25) | 2.18 | " | 1.5 | 4.5 | torn | — | 1.797 | 5897 | 39.73 |
| 41 | " | 0.69 | 2023.4 | Short | Infallible | 0.6715 | .0599 | .007 | 1.95 | (*25) | 2.09 | " | 0.5 | 3.0 | " | — | 1.748 | 5735 | 37.93 |
| 42 | " | 0.95 | 1914.3 | Medium | " | 0.920 | .0551 | .020 | 5.70 | (*20) | 5.83 | " | 1.5 | 7.0 | torn | .326 | 2.055 | 6745 | 40.82 |
| 43 | " | 0.95 | 2020.7 | " | " | 1.2581 | .0582 | .020 | 5.37 | (*25) | 5.76 | " | 1.0 | 6.0 | not torn | .677 | 2.137 | 7011 | 44.14 |
| 45 | " | 1.25 | 2020.7 | " | " | 1.4540 | .0603 | .020 | 11.38 | (*25) | 11.65 | " | 1.5 | 8.0 | torn | .690 | 2.340 | 7680 | 52.93 |
| 47 | " | 1.42 | 3040.5 | " | " | 1.3997 | .0607 | .020 | 6.50 | (*25) | 6.98 | " | 1.5 | 9.0 | not torn | .730 | 2.318 | 7606 | 51.83 |
| 48 | " | 1.42 | 3040.8 | " | " | 1.3997 | .0619 | .020 | 6.03 | (*25) | 6.47 | " | 1.5 | 8.5 | torn | .755 | 2.341 | 7683 | 51.74 |
| 49 | " | 1.42 | 2020.7 | " | " | 1.5200 | .0630 | .030 | 13.00 | (*25) | 13.96 | " | 1.5 | 8.5 | not torn | .730 | 2.257 | 7404 | 43.19 |
| 50 | " | 1.57 | 3039.0 | " | " | 1.5200 | .0630 | .030 | 7.28 | (*25) | 7.82 | " | 1.5 | 9.5 | " | .790 | 2.332 | 7653 | 52.56 |

The length of chamber, l, in the third column, is taken as the distance shown in figure 2 (a).

In the cases of simple harmonic motion in which $d_2$ is not given in the table, the displacements were so large that $d_2$ was prevented from reaching its full extent by the yoke, Y, plate 7, figure 1. Correction for friction was made in these cases by choosing the decrement from some other experiment that would be likely to apply. The number of this experiment is written in parenthesis, in the table, in place of $d_2$. The same procedure is followed in the experiments with direct lift.

Of the experiments in the cylindrical tank, 15 and 16 were performed with the elbow E, figure 17, at the lower end of the 3-inch pipe; no. 17 was performed with this elbow also in place, with the addition of a sheet-iron sleeve in the pipe, to decrease the curvature *at the elbow;* nos. 18 and 19 were performed with the tank empty; and the remaining experiments were performed with the fencing, already described, in position.

The tissue paper was usually torn at one end, and not torn completely off. It was only torn completely off, with small charges, in the experiments with the cylindrical tank *empty* (nos. 18 and 19). The tissue paper was cut one-third across at each end, as already explained, in experiments 15 and 33, inclusive.

The direct-lift impulse-meter was used in experiments 15 to 26, inclusive. In cases in which there was impact of the chamber against the yoke, or pins, at the lower ends of the rods, R, R, plate 6, figure 2, this impulse-meter was useless because of the jar. Only in experiments 16 and 22 was there a measurable displacement, the negligible displacements in the other cases being doubtless due to friction. The spring impulse-meter was used only in the last six vacuum experiments.

An inspection of tables II and III will show that the results, under the same conditions, are in sufficiently close agreement to warrant the comparison of results obtained under various circumstances of firing.

### DISCUSSION OF RESULTS

1. There is a general tendency for the velocities *in vacuo* to be larger than those in air, for the same length of chamber, l, and the same mass of powder.

With Du Pont powder, the medium and short nozzles give greater velocities *in vacuo*. The long nozzle, however, does not show results very much different from those obtained in air.

There is a large difference, however, with Infallible powder, with all three nozzles. For the medium nozzle a comparison of experiments 4 to 12, inclusive, with 35 and 36 shows that the increase amounts to 22 per cent of the velocity in air.

2. The medium nozzle gives, in general, greater velocities than the short or the long nozzle with the same length of chamber, l, and approximately the same charges of powder. In all cases, the short nozzle gives less velocity than the medium or the long nozzle, which is to be expected.

3. The results show no appreciable dependence of the velocities upon the pressure in the tank between 7.5 mm. and 0.5 mm., and it is safe to conclude that the velocities are practically the same from atmospheric pressure down to zero pressure, except as regards the slight increase of velocity with decreasing pressure already mentioned.

4. A comparison of the results when the chamber moved under the influence of the spring with those in which the chamber was merely lifted, show that the agreement of results obtained by the two methods is good, provided the displacement in the direct lift experiment is small (compare experiments 40 and 41 with 26). If, on the other hand, the displacement in the direct lift experiment is *large*, this method gives considerably less velocities than the spring method A comparison of experiments 35 and 36 with 34, 37 and 43 makes it evident that *all the velocities obtained by experiments in which the lift exceeded 4 cm. are from 300 to 600 ft./sec. too small.* This is a very important conclusion, for it means that the highest velocities *in vacuo,* recorded in table III, are *doubtless considerably less than those which were actually attained.*

5. A comparison of the results obtained by means of the circular tank with those obtained by means of the cylindrical tank shows that the velocities range about 100 ft./sec. higher for the circular tank— a difference that is so small as to be well within the accidental variations of the experiments.

Concerning the behavior of the cylindrical tank under different conditions, a comparison of experiments shows that the velocities are much the same for all cases. Hence it is safe to conclude that the rebound, at least for small charges, is not excessive even if an empty tank is used, providing it is sufficiently large.

A check of some interest, on the effectiveness of the cylindrical tank, with the retarder, J, J', in position inside, was the sound of the shot, which resembled a sharp blow of a hammer on the *lower* cap of the 12-inch pipe. The impact was most clearly discernible when the hand was on the lowest part of the tank. The sound, in the case of the circular tank, did not appear to come from any particular part. When the tank was grasped during firing, a throb of the entire tank was noticed.

6. Concerning the proportion of the measured reaction that is due to gaseous rebound, the tissue paper detector, as has already been explained, does not give any information. All that this detector really shows is that the *force* exerted by the initial upward rush of gas past the chamber is not excessive. The fact that the tissue paper is sometimes torn and sometimes not under identical conditions of firing, shows that either this force differs more or less in various parts of the tank (*i. e.,* the upward rush of gas is not perfectly homogeneous) ; or that the tissue paper is weakened by each successive shot. This last explanation is the more probable; for fine particles of the wadding rush upward with the gas, as is proved by fine markings on the smoked glass, and also from the fact that, after a number of shots, the tissue paper is found to be perforated with very small holes.

The gaseous rebound could not be measured accurately with the direct-lift impulse-meter. Thus of all the experiments in which this meter could be used, 15 to 26 inclusive, only two, 16 and 22, gave readable displacements, the failure to obtain readable displacements in the other cases being doubtless due to friction, as already mentioned. It will be noticed that the impulse is under one per cent.

The spring impulse-meter used in the last five experiments gave reliable results because of the very slight friction during operation. This impulse-meter shows that, if the momentum of the chamber were to be corrected for gaseous rebound, *this correction would be much less than one per cent of the momentum of the chamber.* But as has been stated above, the impulse of the rebound at the chamber must be less than that at the impulse-meter, from the fact that gases may pass readily behind the chamber, and exert a downward pressure, and also because of friction in the 3-inch pipe. The effect of gaseous rebound is therefore negligible, and no account of it has been taken in calculating the velocities and efficiencies.

It now becomes possible to find, from the experimental results, the highest velocity *in vacuo* upon which dependence may be placed. This is evidently the result of experiment 45 and is 2.34 km./sec. or 7,680 ft./sec. It is well worth noticing, however, that experiment 50 would have given, without doubt, a velocity even higher, had friction properly been taken into account.

## DISCUSSION OF POSSIBLE EXPLANATIONS

1. The fact that the velocities are higher *in vacuo* than in air seems explicable only by there being conditions of ignition different *in vacuo* from those in air; although this may also have been due to the air in the nozzle interfering with the stream-lines of the gas, thus producing a jet not strictly unidirectional. It should be remarked that the highest velocity *in vacuo* recorded, experiment 25, may have been due to unusually good circumstances of ignition; but may also have been due, in part, to being performed in the circular tank.

2. The fact that the medium nozzle gives in general velocities higher than the long nozzle shows that very likely after traveling the distance from the throat equal approximately to the length of the medium nozzle, the gas is moving so rapidly that it fails to expand fast enough to fill the cross-section of the nozzle. A discontinuity in flow is produced at the place where the gas leaves the wall of the nozzle, and this produces eddying and a consequent loss of *undirectional velocity*. The efficiency could doubtless be increased by constructing the nozzle in the form of a straight portion, corresponding to a cone of 8° taper, for the length of the medium nozzle, with the section beyond this point in the form of a curve concave to the axis of the nozzle.

## CONCLUSIONS FROM EXPERIMENTS

1. The experiments in air and *in vacuo* prove what was suggested by the photographs of the flash in air, namely, that the phenomenon is really a jet of gas having an extremely high velocity, and is not merely an effect of reaction against the air.

2. The velocity attainable depends to a certain extent upon the manner of loading, upon the circumstances of ignition, and upon the form of the nozzle. Hence, in practice, care should be taken to design the cartridge and the nozzle for the density of air at which they are to be used, and to test them in an atmosphere of this particular density.

It is with pleasure that the writer acknowledges the use, as honorary fellow in physics, of the laboratory facilities, and especially the rotary pump, at the Physics Laboratory at Clark University where these experiments were performed.

## SIGNIFICANCE OF THE ABOVE EXPERIMENTS AS REGARDS CONSTRUCTING A PRACTICAL APPARATUS

It will be well to dwell at some length upon the significance of the above experiments. In the first place, the lifting power of both powders is remarkable. Experiment 51 shows, for example, that 42 lbs. can be raised 2 inches by the reaction from less than 0.018 lbs. of powder. One interesting result is the very high efficiency of the apparatus considered as a heat engine. It exceeds, by a wide margin, the highest efficiency for a heat engine so far attained—the " net efficiency " or duty of the Diesel (internal combustion) engine being about 40 per cent, and that for the best reciprocating steam engine but 21 per cent. This high efficiency is, of course, the result of three things: the absence of much heat loss due to the suddenness of the explosion; the almost entire absence of friction; and the high temperature of burning. Owing to these features, it is doubtful if even the most perfect turbine or reciprocating engine could compete successfully with the type of heat engine under consideration.

It is, however, the velocity ($c$ in equations (6) and (7)) which is of the most interest. The highest velocity obtained in the present experiments is 13 ft./sec. under 8,000 ft./sec., thus exceeding a mile and a half per second (the " Parabolic velocity " at the surface of the moon), and also exceeding anything hitherto attained except with minute quantities of matter by means of electrical discharges in vacuum tubes. Inasmuch as the higher velocities range between seven and eightfold that of the Coston rocket we should expect a reduction of initial masses to be made possible by employment of the steel chamber, to at least the *seventh root* of the masses necessary for a chamber like the Coston rocket.

The supposition is, of course, that the mass of propellant material can be made so large in comparison with the mass of the steel chamber, that the latter is comparatively negligible. No attempt was made in the present experiments to reduce the chamber to its minimum weight; in fact, the more massive it was, the more satisfactory could the ballistic experiments be performed. The minimum weight possible, for the same thickness of wall as in the experiments, was calculated by estimating, first, the volume of a chamber from which all superfluous metal had been removed, as shown by the full lines in figure 12, and then calculating the mass of this reduced chamber, from the measured density of the steel. The minimum masses of chamber per gram of powder plus wadding, estimated in this way, were 143, 130, and 120 grams, respectively, for experiments 50, 51, and 52. In the last two cases, a smaller breech-block could doubt-less have been used, as evident from figure 12; and in the first two cases, the chamber wall, itself, could safely have been reduced in thickness. More important still, a " built-up " construction would much reduce the mass as has already been explained.[13]

It should be mentioned that, for any particular chamber, it will be necessary to determine the maximum possible powder charge to a nicety, from the fact that, as modern rifle practice has demonstrated, one charge of dense smokeless powder may be perfectly safe for any number of shots, whereas a slightly larger amount, or the same amount slightly more compressed (a state in which the powder must exist in the present chamber) will result in very dangerous pressures.

But the whole question of ratio of mass-of-powder-to-chamber is without doubt relatively unimportant for the following reason: The photographs of the flash, in experiments 9 and 11, in which the flash was accidentally reflected in the nozzle of the gun, show the nozzle appearing stationary in the photograph, thus demonstrating that the duration of the flash is very small; but this, as already explained, is much longer than the time during which the gases are leaving the nozzle. The time of firing is, therefore, extremely short. This is to be expected, inasmuch as the high pressure in the chamber sets in motion only the small mass of gas and wadding, and hence must exist for a much shorter time than the pressure in a rifle or pistol. For this reason the heat such as is developed in the machine-gun, due to the hot gases remaining in the barrel for an appreciable time during each shot, as well as that due to the friction of the bullet, will be absent in the type of rapid-fire mechanism under discussion. Hence a large number of shots, equivalent to a mass of powder greatly exceeding that of the chamber, may be fired in rapid succession, without serious heating.[14]

## PART III. CALCULATIONS BASED ON THEORY AND EXPERIMENT

### APPLICATION OF APPROXIMATE METHOD

As already explained this method consists in employing the equations

$$M = \frac{R}{a+g} \left( e^{\frac{a+g}{c(1-k)} t} - 1 \right) + e^{\frac{a+g}{c(1-k)} t}, \qquad (6)$$

and

$$M = e^{\frac{at}{c(1-k)}}, \qquad (7)$$

to obtain a minimum M in each interval, where

M = the initial mass, for the interval, when the final mass is one pound, and

R = the air resistance in poundals over the cross-section S, at the altitude of the rocket. If we call, P, the air resistance per unit cross-section, we shall have for R, PS $\frac{\rho}{\rho_0}$ where $\rho$ is the density at the altitude of the rocket, and $\rho_0$ is the density at sea-level.

a = the acceleration in ft. per second$^2$, taken constant throughout the interval,

g = the acceleration of gravity,

t = the time of ascent through the interval, and

c(1−k) = what will be called the " effective velocity," for the reason that the problem would remain unchanged if the rocket were considered to be composed *entirely* of propellant material, ejected with the velocity, c(1−k). It will be remembered that c actually stands for the true velocity of ejection of the propellant, and k for the fraction of the entire mass that consists of material other than propellant. The effective velocity is taken constant throughout any one calculation.

The altitude is divided into intervals short enough to justify the quantities involved in the above equations being taken as constants. The equations are then used to find the minimum value of M for each interval—the mean values of R and g, in the interval, being employed —and the " total initial mass " required to raise a final mass of one pound to a desired altitude is then obtained as the product of these M's.

## VALUES OF THE QUANTITIES OCCURRING IN THE EQUATIONS

*The effective velocity, c(1−k).*—The calculation which follows has been carried out with the assumption [15] of a velocity of ejection of 7,500 ft./sec. and a constant, k, equal to $\frac{1}{15}$. This velocity is considerably less than those that were actually obtained, both in air and *in vacuo.* The " effective velocity " will thus be

$$c(1-k) = 7{,}000 \text{ ft./sec.}$$

It should be noticed that k could be $\frac{1}{12}$ and yet not necessitate a larger velocity of ejection than 7,640 ft./sec., which is also under the highest velocities obtained in the experiments. It is important at this point to remember that the velocities *in vacuo* would doubtless have been found to be considerably higher than the above value, if friction could have been eliminated in the " direct-lift " method.

*The quantity, R.*—The mean value of R for any interval is most easily obtained from a graphical representation of P as a function of v, the mean value of P between the beginning and end of the interval being taken. Three curves have been used for this purpose: for velocities ranging from zero to 1,000 ft./sec., 1,000 to 3,000 ft./sec., and from 3,000 ft./sec. upward. The first curve represented the experimental results of A. Frank[1] obtained with prolate ellipsoids. The second curve represented the experimental results of A. Mallock,[2] whereas the third curve represented an empirical formula by Mallock,[3] which agrees well with experimental results up to 4,500 ft./sec.—the highest velocity that has been attained by projectiles—and hence may be used for still higher velocities with a fair degree of safety. Mallock's expression, reduced to the absolute ft. lb. sec. system and multiplied by $\frac{1}{4}$, the coefficient for projectiles with pointed heads, becomes

$$P = 0.00006432v^2 \left(\frac{v'}{a}\right)^{0.375} + 480 \tag{8}$$

where v′ = the velocity with which a wave is propagated in the air immediately in front of the projectile ; which equals the velocity of the body when that velocity exceeds the velocity of sound in the undisturbed gas ; and

a = the velocity of sound in the undisturbed gas.

The constant, 480 poundals, must be added for velocities over 2,400 ft./sec. owing to the vacuum in the rear of the projectile.

---

[1] A. Frank, Zeitschr. Verein Deutsches Ing. 50, pp. 593-612, 1906.
[2] A. Mallock, Proc. Roy. Soc., 79A, pp. 262-273, 1907.
[3] A. Mallock, Proc. Roy. Soc., 79A, p. 267, 1907.

*The quantity, ρ.*—The above expression (8), for the resistance, holds only at atmospheric pressure. At high altitudes the pressure, of course, decreases greatly. If we call $\rho$ the mean density throughout any interval of altitude, and $\rho_0$ the density at sea-level, the right member of (8), on being multiplied by S and $\frac{\rho}{\rho_0}$ , will give the air resistance, R, experienced by the rocket.

A curve representing the relation between density and altitude up to 120,000 ft. is shown in figure 6. This curve is derived from

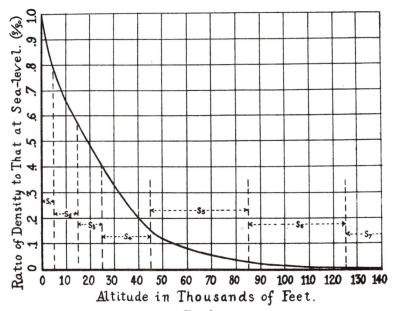

FIG. 6.

a table of pressures and temperatures in Arrhenius' " Lehrbuch der Kosmischen Physik." The ordinates of the curve are the numbers $\frac{\rho}{\rho_0}$ .

Beyond 120,000 ft. the density is calculated by the empirical rule which assumes the density to become halved at every increase in altitude of 3.5 miles. A comparison was made between the values obtained in this way and those obtained from the very probable pressures deduced by Wegener, in the following way: The mean density between two levels for which Wegener gives pressures was obtained by multiplying the difference in pressure by 13.6, and dividing by the difference in level in cm. A comparison showed that the densities used in the present calculations beyond 125,000 ft. were from three to twentyfold larger than those derived from Wegener's data, so that the values used in the present case were doubtless perfectly safe.

Densities beyond 700,000 ft. within the geocoronium sphere, must be negligible, for not only is the density very small but the *resistance to motion* is very small—due, according to Wegener, to the properties of geocoronium—a conclusion which is supported by the fact that meteors remain, for the most part, invisible above this level.

## DIVISION OF THE ALTITUDE INTO INTERVALS

In dividing the altitude into intervals the only condition that must be fulfilled is that the densities in any interval shall not differ widely from the mean value in the interval. The least number of intervals which satisfy this condition are given in the following table:

TABLE IV

| Interval | Length of interval | Height of upper end of interval above sea-level | Mean density in terms of $\rho_0$ | Mean gravity chosen, in terms of gravity at sea-level |
|---|---|---|---|---|
| $s_1$ | 5,000 ft. | 5,000 ft. | 0.928 | 1 |
| $s_2$ | 10,000 | 15,000 | 0.730 | 1 |
| $s_3$ | 10,000 | 25,000 | 0.520 | 1 |
| $s_4$ | 20,000 | 45,000 | 0.278 | 1 |
| $s_5$ | 40,000 | 85,000 | 0.080 | 1 |
| $s_6$ | 40,000 | 125,000 | 0.015 | 1 |
| $s_7$ | 75,000 | 200,000 | 0.0026 | 1 |
| $s_8$ | 300,000 | 500,000 | 0.000025 | 1 |
| $s_{9(a=150)}$ | 3,415,000 | 3,915,000 | . . . . . . . . | 0.839 |
| $s_{9(a=50)}$ | 8,810,000 | 9,310,000 | . . . . . . . | 0.684 |

The mean densities in intervals $s_1$ to $s_6$, inclusive, were obtained from figure 21, on which these intervals are marked. The remaining densities were estimated as already explained.

## CALCULATION OF MINIMUM MASS FOR EACH INTERVAL

The tables V and VI are calculated for a start, respectively, from sea-level and from an altitude 15,000 ft.—*i. e.*, the beginning of $s_3$. The procedure in each case is, however, identical.

The process of calculation is as follows: At the beginning of any interval we have the velocity already acquired during the previous intervals, let us say $v_0$. This velocity is, of course, zero at the beginning of the first interval. Assume any final velocity at random, $v_1$, for the interval in question.

The value of at may be had from the equation

$$v_1 = v_0 + at, \tag{9}$$

and t is at once obtained from the relation

$$s = v_0 t + \tfrac{1}{2} a t^2 \, ; \; i. \; e.,$$

$$t = \frac{s}{v_0 + \tfrac{1}{2} at}, \tag{10}$$

whence, of course, a is at once known.

The calculation of $e^{\frac{a+g}{c(1-k)} t}$ and $e^{\frac{at}{c(1-k)}}$ call for no comment; and R is obtained as P, the mean ordinate between $v_0$ and $v_1$ from the curves as already explained, multiplied by S and $\frac{\rho}{\rho_0}$.

The value of M, the initial mass, for the interval, necessary in order that the final mass in the interval shall be one pound, is then obtained from equation (7); and finally, the ratio of equations (6) to (7) $\left( i. \; e., \; \dfrac{M}{\dfrac{at}{e^{c(1-k)}}} \right)$ is calculated. This is the ratio of the initial mass necessary, including losses due to both R and g, to the mass necessary to give the one pound the same velocity, $v_1$, *without* overcoming R and g; *and the entire calculation must be repeated until a minimum value of this ratio is obtained*—when the corresponding mass, M, will be the minimum mass for the interval in question. Each minimum M is marked in the tables by an asterisk.

This process is carried out for each interval beginning with the first.

It should be noticed that, although P and the density are not really constant in any interval, the result obtained by taking the mean of the quantities must nevertheless give results close to the truth, owing to the fact that P increases during the ascent, whereas the density decreases.

## EXPLANATION OF TABLES V AND VI

It should first be explained why no minimum M has been calculated for the intervals $s_7$ and $s_8$. Although the minima for the preceding intervals are clearly defined, a trial will show that a minimum M can occur, for $s_7$ and $s_8$, only for extremely high velocities, $v_1$; although for $s_7$, a secondary minimum occurs for $v_1 = 8,000$ ft./sec. Even for $v_1 = 30,000$ ft./sec. the minimum has not yet been attained for this interval, although the acceleration required to produce this velocity

TABLE V

| Interval | $v_i$ feet/sec. | $at$ | $t$ sec. | $a$ | $\frac{ut}{c(1-x)}$ | $\frac{a-y}{c(1-x)}t$ | $e^{\frac{at}{c(1-x)}}$ | $e^{\frac{(a-y)t}{c(1-x)}}$ | $P$, pounds per sq. in. | $R$, (PS%) | $\frac{R}{a+g}$ | $M$ lbs. | $\frac{M}{e^{\frac{at}{c(1-x)}}}$ | $M_z$ lbs. | $e^{\frac{z\,at}{c(1-x)}}$ | $M_{R1}$ lbs. | $e^{\frac{g^2(a-y)t}{c(1-x)}}$ | $M_{R2}$ lbs. | Time to upper end of Interval |
|---|---|---|---|---|---|---|---|---|---|---|---|---|---|---|---|---|---|---|---|
| $s_1$ * | 500 | 500 | 20.0 | 25 | .0716 | .1630 | 1.074 | 1.176 | 7.36 | 6.85 | .120 | 1.1972 | 1.113 | | | | | | |
| | 800 | 800 | 12.5 | 64 | .1145 | .1720 | 1.120 | 1.186 | 20.0 | 18.5 | .193 | 1.2218 | 1.092 | 1.1584 | 1.458 | 4.586 | 167.3 | 2 03.90 | 10.0 sec. |
| | 1000 | 1000 | 10.0 | 100 | .143 | .1890 | 1.153 | 1.207 | 31.25 | 29.0 | .219 | 1.252 | 1.086 | | | | | | |
| | 1200 | 1200 | 8.34 | 144 | .172 | .212 | 1.185 | 1.235 | 61.4 | 57.0 | .323 | 1.311 | 1.106 | | | | | | |
| | 1500 | 1500 | 6.7 | 226 | .215 | .2475 | 1.242 | 1.276 | 104.6 | 98.0 | .378 | 1.380 | 1.112 | | | | | | |
| | 2000 | 2000 | 5.0 | 400 | .267 | .309 | 1.332 | 1.362 | 202.5 | 188.0 | .436 | 1.6195 | 1.138 | | | | | | |
| $s_2$ * | 1100 | 100 | 9.54 | 10.47 | .0143 | .0578 | 1.014 | 1.061 | 153.3 | 112.1 | 2.64 | 1.222 | 1.206 | 1.4860 | 1.150 | 3.155 | 6.73 | 20.60 | 19.1 |
| | 1200 | 200 | 9.1 | 22.0 | .0286 | .0704 | 1.034 | 1.073 | 166.6 | 121.6 | 2.24 | 1.237 | 1.199 | | | | | | |
| | 1400 | 400 | 8.33 | 47.9 | .0574 | .0954 | 1.060 | 1.100 | 216.0 | 158.7 | 1.97 | 1.297 | 1.223 | | | | | | |
| $s_3$ * | 1300 | 100 | 8.0 | 12.5 | .0143 | .0508 | 1.014 | 1.052 | 2.50.0 | 130.0 | 2.925 | 1.204 | 1.186 | 1.462 | 1.137 | 2.974 | 5.62 | 16.52 | 26.8 |
| | 1400 | 200 | 7.7 | 25.8 | .0286 | .0637 | 1.034 | 1.066 | 262.8 | 136.9 | 2.37 | 1.222 | 1.182 | | | | | | |
| | 1600 | 400 | 7.15 | 56.4 | .0574 | .0906 | 1.06 | 1.096 | 294.5 | 152.6 | 1.74 | 1.261 | 1.191 | | | | | | |
| $s_4$ * | 1500 | 100 | 13.8 | 7.23 | .0143 | .0775 | 1.014 | 1.080 | 339.0 | 94.3 | 2.42 | 1.273 | 1.255 | 1.626 | 1.198 | 3.91 | 11.33 | 33.73 | 40.13 |
| | 1600 | 200 | 13.33 | 15.0 | .0286 | .0898 | 1.034 | 1.094 | 372.0 | 101.5 | 2.17 | 1.297 | 1.253 | | | | | | |
| | 1700 | 300 | 12.9 | 23.24 | .0429 | .1022 | 1.046 | 1.107 | 394.0 | 109.4 | 1.975 | 1.319 | 1.26 | | | | | | |
| | 1800 | 400 | 12.5 | 33.25 | .0574 | .1170 | 1.060 | 1.123 | 424.0 | 118.0 | 1.81 | 1.346 | 1.267 | | | | | | |
| $s_5$ * | 1700 | 100 | 24.25 | 4.125 | .0143 | .1258 | 1.014 | 1.133 | 439.0 | 35.1 | .974 | 1.262 | 1.245 | 1.711 | 1.313 | 4.304 | 40.70 | 88.45 | 63.83 |
| | 1800 | 200 | 23.7 | 8.45 | .0286 | .1566 | 1.034 | 1.146 | 480.0 | 38.4 | .951 | 1.2845 | 1.242 | | | | | | |
| | 2000 | 400 | 22.24 | 18.0 | .0574 | .159 | 1.06 | 1.173 | 535.0 | 42.8 | .854 | 1.321 | 1.246 | | | | | | |
| $s_6$ * | 1900 | 100 | 21.7 | 4.62 | .0143 | .1135 | 1.014 | 1.12 | 567. | 8.50 | .232 | 1.1478 | 1.13 | 1.3406 | 1.280 | 2.810 | 29.76 | 36.02 | 84.93 |
| | 2000 | 200 | 21.1 | 9.50 | .0286 | .1255 | 1.034 | 1.133 | 603. | 9.01 | .2175 | 1.162 | 1.123 | | | | | | |
| | 2200 | 400 | 20.0 | 20.0 | .0574 | .1490 | 1.06 | 1.16 | 669. | 10.02 | .1923 | 1.1907 | 1.124 | | | | | | |
| $s_7$ (a=150) | 5160 | 3160 | 21.0 | 150 | .4523 | .5452 | 1.572 | 1.725 | 1878. | 4.84 | .0264 | 1.7442 | 1.108 | 3.022 | 2.97 | 53.96 | $2.63{\times}10^5$ | $2.70{\times}10^6$ | 105.93 |
| (a=50) | 3393 | 1393 | 27.8 | 50 | .199 | .3276 | 1.218 | 1.387 | 1122. | 3.1 | .0355 | 1.4007 | 1.15 | 1.9319 | 1.900 | 11.13 | $7.03{\times}10^3$ | $7.28{\times}10^3$ | 112.73 |
| $s_8$ (a=150) | 10790 | 5630 | 37.5 | 150 | .804 | .976 | 2.23 | 2.65 | 10600. | 0.272 | .00146 | 2.6524 | 1.19 | 7.0288 | 7.02 | 1193.7 | $2.88{\times}10^6$ | $2.88{\times}10^{10}$ | 143.43 |
| (a=50) | 6833 | 2840 | 55.8 | 50 | .399 | .652 | 1.49 | 1.92 | 4000. | 0.0994 | .00121 | 1.9211 | 1.293 | 3.6832 | 3.680 | 117.54 | $4.67{\times}10^7$ | $4.67{\times}10^7$ | 168.53 |
| $s_9$ (a=150) | 33790 | 23000 | 153.5 | 150 | 3.29 | 3.89 | 26.9 | 48.8 | | | | 48.8 | | 23800 | 23800.0 | $1.906{\times}10^{12}$ | $5.74{\times}10^{16}$ | $5.74{\times}10^{16}$ | 296.93 |
| (a=50) | 30533 | 23700 | 472.5 | 50 | 3.38 | 4.85 | 29.13 | 129.0 | | | | 129.0 | | 16700 | 16700.0 | $1.995{\times}10^{15}$ | $1.25{\times}10^{17}$ | $1.25{\times}10^{17}$ | 641.03 |

is 6,000 ft./sec.² The reason for this state of affairs is evident at once from the fact that the density ratio, $\frac{\rho}{\rho_0}$ is very small for $s_7$, and also from the fact that a occurs in the denominator of the term containing R in equation (6), so that the large acceleration counterbalances the increase in R.

Thus, in order that the initial mass for $s_7$ shall be a minimum, the acceleration must become very large, with consequent severe strains in the rocket apparatus and instruments carried by the rocket, to say nothing of the difficulty of firing with sufficient rapidity to produce such large accelerations. It thus becomes advisable to choose a moderate acceleration in $s_7$ and $s_8$, and not to assign a velocity, $v_1$, as was done in the preceding intervals. Two accelerations are chosen: 50 ft./sec.² and 150 ft./sec.², respectively. The interval $s_9$, also calculated for assigned accelerations, will be explained in detail below. In all cases, when either one of these accelerations is mentioned in connection with $s_8$ and $s_9$, this acceleration will be understood as having been taken also in the preceding intervals, beyond $s_6$.

In order to see how far the effective velocity, $c(1-k)$ may fall short of 7,000 ft./sec. and still not render the rocket impracticable, a few additional columns for M are calculated.

In the first of the additional columns, $M_2$, the effective velocity is taken as 3,500 ft./sec., namely, half that of the preceding calculations. This allows of considerable inefficiency of the apparatus, in a number of ways. For example, the product

$$c(1-k) = 3,500,$$

may be given by the same proportionality, k, as before, but with a velocity of ejection of the gases as low as 3,750 ft./sec. On the other hand, the velocity of ejection may be as large as before (*i. e.*, 7,500 ft./sec.); and the proportionality, k, increased to 0.533; meaning, of course, that the rocket now consists more of mechanism than of propellant.

The second additional calculations, $M_{R_1}$, are carried out under the assumption that a reloading mechanism is used, with k as in the original calculations ($k=\frac{1}{15}$), but that the velocity of expulsion of the gases is the mean found by experiment for the Coston ship rockets, namely 1029.25 ft./sec. In this case the effective velocity is

$$c(1-k) = 1029.25(1-\tfrac{1}{15}) = 960 \text{ ft./sec.}$$

Table VI

| Interval | $v$, ft./sec | $at$ | $t$, sec | $a$ | $\frac{at}{c(1-x)}$ | $\frac{a+g}{c(1-x)}t$ | $e^{\frac{at}{c(1-x)}}$ | $e^{\frac{(a+g)t}{c(1-x)}}$ | $P$, poundals per sq.in. | $R$, $(PS\%/P_B)$ | $\frac{R}{a+g}$ | $M$, lbs. | $\frac{M}{e^{\frac{at}{c(1-x)}}}$ | $e^{\frac{2(a+g)t}{c(1-x)}}$ | $M_2$, lbs. | $e^{\frac{728(a+g)t}{c(1-x)}}$ | $M_{R1}$, lbs. |
|---|---|---|---|---|---|---|---|---|---|---|---|---|---|---|---|---|---|
| $s_3$ | 500 | 500 | 40. | 12.5 | .0715 | .255 | 1.074 | 1.29 | 11.53 | 5.97 | .134 | 1.329 | 1.236 | 1.574 | 1.718 | 5.225 | 6.545 |
| * | 800 | 800 | 25. | 32.0 | .1147 | .2277 | 1.120 | 1.256 | 30.7 | 16.00 | .250 | 1.300 | 1.162 | | | | |
|  | 1000 | 1000 | 20. | 50.0 | .142 | .235 | 1.152 | 1.263 | 46.7 | 24.3 | .295 | 1.341 | 1.165 | | | | |
|  | 1500 | 1500 | 13.4 | 112.0 | .2145 | .277 | 1.24 | 1.318 | 165.0 | 83.3 | .570 | 1.499 | 1.207 | | | | |
| $s_4$ | 900 | 100 | 23.7 | 4.23 | .0143 | .1227 | 1.013 | 1.132 | 957 | 27.7 | .764 | 1.232 | 1.216 | 1.293 | 1.518 | 2.581 | 3.794 |
| * | 1000 | 200 | 22.2 | 9.00 | .0286 | .1305 | 1.034 | 1.157 | 108.8 | 31.4 | .767 | 1.242 | 1.200 | | | | |
|  | 1300 | 500 | 19.1 | 26.2 | .0714 | .1645 | 1.073 | 1.177 | 165.0 | 46.25 | .794 | 1.318 | 1.227 | | | | |
|  | 1800 | 1000 | 15.4 | 65.0 | .1430 | .2136 | 1.152 | 1.238 | 305.0 | 87.90 | .908 | 1.455 | 1.263 | | | | |
| $s_5$ | 1100 | 100 | 38.1 | 2.625 | .0124 | .1888 | 1.013 | 1.207 | 150.1 | 12.0 | .347 | 1.278 | 1.261 | 1.495 | 1.685 | 4.32 | 5.594 |
|  | 1200 | 200 | 36.5 | 5.47 | .0286 | .1960 | 1.03 | 1.215 | 170.0 | 13.55 | .362 | 1.293 | 1.255 | | | | |
| * | 1300 | 300 | 34.75 | 8.64 | .0430 | .202 | 1.044 | 1.223 | 195.0 | 15.65 | .384 | 1.306 | 1.250 | | | | |
|  | 1400 | 400 | 33.3 | 12.0 | .0571 | .210 | 1.058 | 1.233 | 218.0 | 17.49 | .397 | 1.325 | 1.252 | | | | |
|  | 1500 | 500 | 32.1 | 15.60 | .0715 | .2192 | 1.073 | 1.245 | 243.5 | 19.45 | .520 | 1.372 | 1.280 | | | | |
|  | 2000 | 1000 | 26.1 | 21.40 | .1147 | .268 | 1.12 | 1.308 | 417.0 | 33.4 | .623 | 1.501 | 1.340 | | | | |
| $s_6$ | 1600 | 300 | 27.7 | 10.8 | .0430 | .1690 | 1.045 | 1.184 | 343.0 | 5.16 | .1203 | 1.206 | 1.153 | 1.522 | 1.581 | 4.66 | 5.075 |
|  | 1800 | 500 | 25.7 | 19.5 | .0714 | .1890 | 1.074 | 1.206 | 406.0 | 6.10 | .1186 | 1.230 | 1.147 | | | | |
|  | 1900 | 600 | 25.0 | 24.0 | .0857 | .201 | 1.091 | 1.223 | 430.0 | 6.43 | .1150 | 1.248 | 1.147 | | | | |
| * | 2000 | 700 | 24.2 | 28.9 | .1002 | .212 | 1.105 | 1.234 | 460.0 | 6.90 | .1134 | 1.260 | 1.140 | | | | |
|  | 2100 | 800 | 23.6 | 33.8 | .1142 | .224 | 1.118 | 1.249 | 510.0 | 7.65 | .1165 | 1.278 | 1.142 | | | | |
|  | 2200 | 900 | 22.8 | 40.0 | .1285 | .237 | 1.124 | 1.266 | 534.0 | 8.02 | .1115 | 1.295 | 1.151 | | | | |

The third additional calculations, $M_{R_2}$, are carried out for the case of a rocket built up of Coston rockets in bundles (shown in section in fig. 22), the lowest bundle of which is fired first and then released; after which the bundle above is fired and then released, and so on. For the Coston ship rocket (having a range of a quarter of a mile, with the charge of red fire removed, as already stated) the ratio of the powder charge to the remaining mass of the rocket is found to be closely $\frac{1}{4}$. Hence the " effective velocity " in this case is only

$$c(1-k) = 1029.25(1 - \tfrac{4}{5}) = 257.3 \text{ ft./sec.}$$

The M's in the last two cases are calculated only for the accelerations that make M minima for the first case (effective velocity, 7,500 ft./sec.). Hence in these cases, the M's are not minima, although only in the last two cases is there probably much discrepancy from the actual minima.

The cross-section, *throughout any interval,* is taken as one square inch except for interval $s_9$. It will be seen from the table that this is justifiable, as the largest mass in intervals $s_1$ to $s_8$ does not differ much from one pound.

## CALCULATION OF MINIMUM MASS TO RAISE ONE POUND TO VARIOUS ALTITUDES IN THE ATMOSPHERE

The "total initial masses" required to raise one pound from sea-level to the upper end of intervals $s_6$, $s_7$ and $s_8$ are given in table VII. They are obtained by multiplying together the minimum masses (marked by stars in table V), from $s_1$ up to and including the interval in question, and represent, as already explained, the mass in pounds of a rocket which, starting at sea-level, would become reduced to one pound at the altitude given.

The highest altitude attained by the one pound mass is not, however, the upper end of the interval in question, but is a very considerable distance higher. This, of course, follows from the fact that the one pound reaches the upper end of each interval with a considerable velocity, and will continue to rise after propulsion has ceased until this velocity is reduced to zero, by gravity and air resistance.

If we call $v_n$ the velocity with which the pound mass reaches the upper end of the particular interval where propulsion ceases, h the distance beyond which the one pound will rise (the cross-section still being one square inch), and p the mean air resistance in poundals

TABLE VII

| Interval | Altitude of Upper End of Interval in feet | Greatest Altitude Attained (feet) | Time (sec.) to reach Greatest Altitude from sealevel | Total Initial Masses (in lbs.) for One Pound Final Mass | | | | | | | |
|---|---|---|---|---|---|---|---|---|---|---|---|
| | | | | Starting from | | | Sea-level | | Starting from 15000 feet | | |
| | | | | $c(1-\kappa)=7000$ | $c(1-\kappa)=3500$ | $c(1-\kappa)=960$ | $c(1-\kappa)=257.3$ | $c(1-\kappa)=257.3$ R taken $=0$ | $c(1-\kappa)=7000$ | $c(1-\kappa)=3500$ | $c(1-\kappa)=960$ |
| $s_6$ | 125,000 | 184,500 | 144.13 | 3.665 | 12.61 | 2030.0 | $7.40\times10^9$ | $8.63\times10^8$ | 2.66 | 6.95 | 702.0 |
| $s_7$ (a=50) | 200,000 | 377,500 | 217.73 | 5.14 | 24.36 | $2.26\times10^4$ | $5.46\times10^{14}$ | $6.08\times10^{11}$ | 3.74 | 13.38 | 7820. |
| (a=150) | 200,000 | 610,000 | 265.93 | 6.40 | 38.10 | $1.096\times10^5$ | $2.00\times10^{15}$ | $2.28\times10^{14}$ | 4.65 | 20.90 | 37800. |
| $s_8$ (a=50) | 500,000 | 1,228,000 | 380.53 | 9.875 | 89.60 | $2.66\times10^6$ | $2.55\times10^{19}$ | $2.89\times10^{18}$ | 7.19 | 49.30 | $9.17\times10^5$ |
| (a=150) | 500,000 | 2,310,000 | 475.23 | 12.33 | 267.70 | $1.318\times10^8$ | $5.77\times10^{26}$ | $6.53\times10^{25}$ | 8.97 | 147.30 | $4.51\times10^7$ |
| $s_9$ (a=50) | 9,310,000 | ∞ | ∞ | 1274.0 | $1.497\times10^6$ | $5.32\times10^{21}$ | $3.21\times10^{76}$ | $3.63\times10^{75}$ | 926.0 | $8.22\times10^5$ | $1.82\times10^{21}$ |
| (a=150) | 3,915,000 | ∞ | ∞ | 602.0 | $6.37\times10^5$ | $2.49\times10^{20}$ | $3.32\times10^{71}$ | $3.76\times10^{70}$ | 438.0 | $3.51\times10^5$ | $8.59\times10^{19}$ |

over the distance h, we have by the Principle of Work and Energy,

$$h = \frac{v_n^2}{2(g+p)}.$$

The values of p are small, owing to small atmospheric density, being 1.59 poundals for the h beyond $s_6$; 0.28 beyond $s_7$ (a = 50); and 0.465 beyond $s_7$ (a = 150). For $s_8$ the low density makes this quantity negligible.

The altitudes obtained by adding to the interval the corresponding h, are called the " Greatest altitude attained " in table VII.

Obviously if the start is made at a high elevation, the " total initial mass " required to reach a given height will be less than for a start at sea-level, due not only to the fact that the apparatus is not raised through so great a height, but also to the fact that the denser part of the atmosphere is avoided. Table VI gives minimum masses, M, calculated for a start with zero velocity from the beginning of interval $s_3$ (i. e., 15,000 ft.), the effective velocity being 7,000 ft./sec., as in table V.

It happens that the velocity $v_1$ for minimum M in the interval $s_6$ of table VI is the same as the $v_1$ for the same interval in table V. The calculations that have been made for the intervals beyond $s_6$ apply therefore to the present case, and the only difference between the two cases is that the masses required to reach $s_7$ will be greater, for the start at sea-level, than for the start at 15,000 ft.

The calculations, beginning at 15,000 ft. have been carried out in table VII for all but the lowest " effective velocity "; and it will be observed that the start from a high elevation becomes important only for the lower " effective velocities."

The most striking as well as the most important conclusion to be drawn from table VII is the small " total initial mass " required to raise one pound to very great altitudes when the " effective velocity " is 7,000 ft./sec., the mass for the height of 437 miles (2,310,000 ft.) for example, being but 12.33 lbs., starting from sea-level. *Even for an " effective velocity " of 3,500 ft./sec.,* which allows of considerable inefficiency in the rocket apparatus, *the mass is sufficiently moderate to render the method perfectly practicable,* for in this case an altitude of over 230 miles from sea-level, practically the limit of the earth's atmosphere, requires under 90 lbs.[16]; and an altitude of 118 miles, close under the geocoronium sphere, only 38 lbs. For a start at 15,000 ft., the masses are of course, less, namely 49.3 lbs. and 20.9 lbs., respectively.[17]

The enormous difference between the total initial masses required for low-efficiency rockets, compared with those for high, may at first appear surprising; but they should be expected from the exponential nature of equations (6) and (7). Thus if the " effective velocity " is reduced from 7,000 ft./sec. to half this value, the mini-

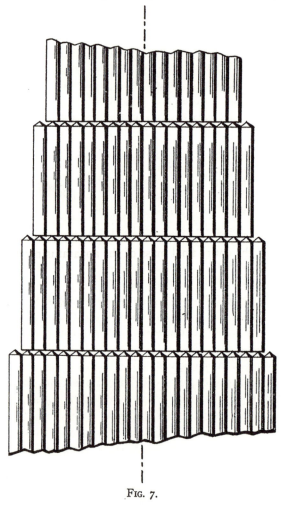

FIG. 7.

mum masses for each interval, neglecting air resistance, will be those for 7,000 ft./sec. *squared;* and including air resistance, still greater. Similarly for an effective velocity of 960 ft./sec. which is that for reloading rockets having the same velocity of ejection as Coston ship rockets, the minimum masses will be those for 7,000 ft./sec. *raised to the 7.28th* power; and for bundles or groups of ship rockets, as shown in figure 7, the minimum masses will be those for 7,000 ft./sec., raised to the *27.2th* power. Even when air resistance is entirely neglected in the calculations for the last case, the masses are of much the same magnitude, as shown in table VII. The large values of the masses $M_{R_1}$ and $M_{R_2}$ simply express the impossibility of employing rockets of low efficiency. Attention may be called to the particular case under $M_{R_2}$ (the groups of ship rockets indicated in fig. 7) in which one pound is raised to the altitude of 1,228,000 feet (232 miles); the " total initial mass " in this case, even neglecting air resistance entirely, is $2.89 \times 10^{18}$ lbs., or *over sixfold greater than the entire mass of the earth.*

These large numbers, to be sure, agree with one's first impression as to the probable initial mass of a rocket designed to reach extreme altitudes; but the comparatively small initial masses, possible with high efficiency, are not intuitively evident until one realizes what an enormous reduction is involved in extracting anything as large as the 27th root of a number.

It should be observed that the apparatus is taken as weighing one pound. Strictly speaking, if the recording instruments have a mass of one pound, the entire final mass of the apparatus must be at least three or four pounds. The mass for the recording instruments may be considered as being very small, yet many valuable researches could, of course, be performed with an apparatus weighing no more than this.[18] The entire final apparatus should if possible be designed to weigh not over 3 or 4 lbs. at most, unless the efficiency of the apparatus is so high that the " effective velocity," $c(1-k)$, is at least in the neighborhood of 7,000 ft./sec. An examination of table VII makes very evident the *necessity of securing maximum effectiveness of the apparatus before a rocket for such a purpose as meteorological work, for example, is constructed; in order to make the method as inexpensive as possible.* It should be remarked, however, that the " total initial mass " *will really not be increased in as large a proportion as the final mass* if the latter is made greater than one pound by virtue of equation (2).

Before proceeding further it will be well to consider carefully the question of air resistance as dependent upon the cross-section of the rocket during flight. It has already been assumed that the cross-section, in the calculation of the minimum M for each interval, was one square inch. If we make the apparatus as long, narrow, and compact, as possible, the assumption of a cross-section of one square inch for an apparatus weighing one pound will not be unreasonable. A glance at tables V and VI will show that, for " effective velocities " of 7,000 ft./sec. and 3,500 ft./sec., the mass at the beginning of any interval (except $s_9$) does not greatly exceed one pound—the mass at the end of each interval being one pound—so that the computations are in agreement with this assumption of area of cross-section. For the two cases of the adapted Coston rockets, the masses at the beginning of the intervals are much larger; and hence we see that the " total initial masses " in table VII, large as they are, would have been even larger if a proper value of cross-section had been employed.

The important point is, however, that cross-sectional areas of *even less than one square inch should have been used*. The reason for this is obvious when one remembers that in calculating the " total initial masses," when we multiply minimum masses, M, together we are also multiplying the cross-sections in the same ratio. In other words, we are considering numbers of rockets, each of one square inch cross-section, grouped together side by side, into a bundle. But such an arrangement would have its cross-section proportional to its *mass* and not to the $\frac{2}{3}$d power of its mass, as would be the case if the *shape of the rocket apparatus were at all times similar to the shape at the start* (as in the ideal rocket, fig. 1). This constant similarity of shape is, as we have seen (equation 2), one of the conditions for a minimum initial mass. Hence the " total initial masses " that have been calculated are *really larger* than the true minima, which would be obtained only by repeating the calculations, assuming a smaller cross-section except in the last few intervals, in which the rocket has become so small that the condition of one-square-inch-per-pound is approximately satisfied.

Before leaving the subject of air resistance, attention should be called to the fact that the velocities (table V) do not exceed that for which air resistance has been studied by Mallock until in $s_7$, for $a = 150$ ft./sec.², and in $s_8$, for $a = 50$ ft./sec.²; and furthermore, that the velocities do not become much in excess until the densities have become almost negligible.

### CHECK ON APPROXIMATE METHOD OF CALCULATION

A simple calculation, involving only the most elementary formulæ instead of equations (6) and (7) will show that the " total initial masses " in table VII cannot be far from the truth.

Consider, for simplicity, a rocket of the form shown in figure 1, and suppose that one-third of the mass of the rocket is fired downward, with a velocity of 7,000 ft./sec. at the first shot; one-third of the remaining mass, at the second shot; and so on, for successive shots. From the principle of the Conservation of Momentum it will be evident that the mass that remains is given an *additional* upward velocity of 3,500 ft./sec. *after each shot.*

Thus, after the fourth shot, the mass that remains is $\frac{16}{81}$, or practically $\frac{1}{5}$, of the initial mass, and the velocity is 14,000 ft./sec. This velocity is sufficient, *if we neglect air resistance,* to raise the part of the rocket that remains to an altitude of *580 miles* (by the familiar relation, $v^2 = 2gh$). Although the range would be much reduced if air resistance were considered, it should nevertheless be remembered that the values in table VII are calculated *for the condition under which air resistance is a minimum.*

The above simple case is not realizable in practice because of the large mass of propellant for each shot compared with the total mass— *i. e.,* provision is not made for the mass of the chamber. The result will be the same, however, if smaller charges are fired in rapid succession, as will be evident from a calculation similar to the above, which is carried out in Appendix E, page 63, under the assumption of smaller charges for successive shots.

## RECOVERY OF APPARATUS ON RETURN

A point of considerable practical importance is the question of finding the apparatus on its return, and of following it during flight, both of which depend in a large measure upon the time of flight.

Concerning the times of ascent, table VII shows that these are remarkably short. For example a height of over 230 miles is reached in less than $6\frac{1}{2}$ minutes ($s_8$; $a = 50$). The reason is, of course, that the rocket under present discussion possesses the advantage of the bullet in attaining a high velocity, with the added advantage of starting gradually from rest. In fact, the motion fulfills closely the ideal conditions for extremely rapid transit—namely, starting from rest with the maximum acceleration possible, and reversing this acceleration, in direction, at the middle of the journey.

The short time of ascent and descent is, of course, highly advantageous as regards following the apparatus during ascent, and recovering it on landing. The path can be followed, by day, by the ejection of smoke at intervals, and at night by flashes. Any distinctive feature, as for example, a long black streamer, could assist in rendering the instruments visible on the return.

Some means will, of course, be necessary to check the velocity of the returning instruments. It might not appear, at first sight, that a parachute would be operative at a velocity of 10,000 ft./sec. or more; but it should be remembered that this velocity will occur in air of very small density, so that the pressure, or force per unit area of the parachute, would not be excessive, notwithstanding the high velocity of the apparatus. The magnitudes of the air resistance will, of course, be much larger than would be indicated from the values of R in tables V and VI, from the fact that, for motion with the parachute, the cross-section will be much larger in proportion to the mass of the rocket than for the cases presented in these tables.

If the parachute is so large that the velocity will be decreased greatly when the denser air is reached, the descent will be so slow that finding of the apparatus will not be so easy as would be the case with a more rapid descent. For this reason, part of the parachute device must be lost automatically when the apparatus has fallen into air of a certain density; or else the parachute must be small enough to facilitate a rapid descent, with additional parachute devices rendered operative as the rocket nears the ground. Such devices are not described in the present paper, but can be of simple and light construction.

The effectiveness of a parachute of even moderate size, operating in a region where the density is small, may be demonstrated by the following concrete example. Suppose that an apparatus weighing one pound and having a parachute of one square foot area descends from the altitude, 1,228,000 ft. (over 200 miles), and does not encounter any atmospheric resistance until it is level with the upper limit of $s_6$ (125,000 ft.). This condition will not, of course, be that which would actually obtain in practice, for a continually increasing resistance will be experienced as the apparatus descends; but if a sufficient braking action can be shown to exist in the present example, the parachute device will *a fortiori* be satisfactory in practice.

The velocity acquired by the apparatus in falling freely under the influence of gravity between the two levels is

$$\sqrt{64 \times 1,103,000} = 8,400 \text{ ft./sec.}$$

Now the air resistance in poundals per square inch of section at atmospheric pressure for this velocity is, from the plot of Mallock's formula, $360 \times 32$ poundals per square inch, making the value of R for the area of the parachute

$$R = 1,653,000 \text{ poundals/in.}^2$$

But the actual resistance is R, multiplied by the relative density at 125,000 ft. which is approximately 0.01, giving for the resistance,

$$F = 16,530 \text{ poundals/in.}^2$$

A retarding acceleration must therefore act upon the apparatus, of amount given by

$$a = \frac{F}{M} = 16,530 \text{ ft./sec.}^2$$

*Hence it is safe to say that, long before the apparatus had fallen to the 125,000 ft. level, the velocity would have been reduced to, and maintained at, a safe value, with the employment of even a small parachute.* This case, it should be noticed, is entirely different from that of a falling meteor; in that the apparatus under discussion falls from rest, at the highest point reached, whereas the meteor enters the earth's atmosphere with an enormous initial velocity.

If it is considered desirable, for any reason, to dispense with a sufficiently large parachute, the retarding of the apparatus may be accomplished to any degree by having the rocket consist, at its highest point of flight, not merely of instruments plus parachute, but of instruments together with a chamber, and considerable propellant material. Then, after the rocket has descended to some lower level, let us say, to the upper limit of $s_6$, this propellant material can be ejected, so that the velocity is considerably checked before the apparatus reaches as low an altitude as, say, 5,000 ft. For the cases in which the effective velocity, $c(1-k)$, is as large as 7,000 ft./sec. there is little inconvenience in increasing the mass in this way. But for the case in which $c(1-k) = 3,500$, this method can hardly be as satisfactory as the parachute method; for if the "final" mass to be elevated is made a number of pounds, let us say n, the "total initial mass" (which is large even for one pound final mass) will be n fold larger, and the apparatus correspondingly more expensive.

## APPLICATIONS TO DAILY OBSERVATIONS

Before leaving the subject of the attainment of high altitudes within the earth's atmosphere, it will be well to mention briefly another application of the method herein discussed: namely, to the sending daily of small recording instruments to *moderate* altitudes, such as five or six miles. As is already understood, simultaneous daily observations of the vertical gradients of pressure, temperature, and wind velocity, at a large number of stations would doubtless be of great value in weather forecasting. The method herein described is evidently well suited for such a purpose, in that the time of rise and fall would be short, so that the apparatus could easily be found on the return. Thus the expense would be slight, being simply that of a fresh magazine of cartridges for each day.

For this work, as well as for that previously described, the head of the rocket should be prevented from rotating, by means of a gyroscope, such as is explained in United States Patent, No. 1,102,653.

## CALCULATION OF MINIMUM MASS REQUIRED TO RAISE ONE POUND TO AN "INFINITE" ALTITUDE

From the fact that the preceding calculation leads us to conclude that such an extreme altitude as 2,310,000 ft. (over 437 miles) can be reached by the employment of a moderate mass, *provided the efficiency is high*, it becomes of interest to speculate as to whether or not a velocity as high as the "parabolic" velocity for the earth could be attained by an apparatus of reasonably small initial mass.

Theoretically, a mass projected from the surface of the earth with a velocity of 6.95 miles/sec. would, *neglecting air resistance,* reach an infinite distance, after an infinite time; or, in short, would never return. Such a projection without air resistance, is, of course, impossible. Moreover, the mass would not reach infinity but would come under the gravitational influence of some other heavenly body.

We may, however, consider the following conceivable case: If a rocket apparatus such as has here been discussed were projected to the upper end of interval $s_8$, either with an acceleration of 50 or 150 ft./sec.$^2$, and *this acceleration were maintained to a sufficient distance beyond $s_8$*, until the parabolic velocity were attained, the mass finally remaining would certainly never return.

If we designate as the upper end of $s_9$ the height at which the velocity of ascent becomes the "parabolic" velocity, it will be evident that this height will be different for the two accelerations chosen, inasmuch as the "parabolic" velocity decreases with increasing distance from the center of the earth.

If we call $u =$ the "parabolic" velocity at a distance H above the surface of the earth,

$v_1 =$ the velocity acquired at the upper end of interval $s_8$,

$s_0 =$ the height of the upper end of $s_8$ above sea-level, ·

we have, taking the radius of the earth as 20,900,000 feet,

$$u = v_1 + at, \tag{11}$$
$$H = s_0 + v_1 t + \tfrac{1}{2}at^2, \tag{12}$$

and also the equation relating " parabolic " velocity to distance from the center of the earth

$$\frac{36,700}{u} = \sqrt{\frac{20,900,000 + H}{20,900,000}}. \tag{13}$$

On putting the values of u and H, from (11) and (12), in (13), we have

$$\sqrt{20,900,000} \times 36,700 = (v_1 + at) \sqrt{21,400,000 + v_1 t + \tfrac{1}{2}at^2}. \tag{14}$$

Equation (14) is a biquadratic in t, from which t may easily be obtained (by trial and error). The values of t, for the two accelerations chosen, given in table V, enables u and the initial masses for $s_9$, to be at once obtained.

The effect of air resistance in $s_9$ is negligible, if we accept Wegener's conclusions, above mentioned, concerning the properties of geocoronium. But even if we use the empirical rule of a fall of density to one-half for every 3.5 miles we shall find the reduction of velocity very small on passing from the upper end of $s_8$ (500,000 ft.) to 1,000,000 ft. (beyond which the density is negligible). This is shown in Appendix F, page 64.

The " total initial masses," to raise one pound to an " infinite " altitude, for the two accelerations chosen, are given in table VII. It will be observed that they are astonishingly small, *provided the efficiency is high.* Thus with an " effective velocity " of 7,000 ft./sec., and an acceleration of 150 ft./sec.$^2$, the " total initial mass," starting at sea-level is 602 lbs., and starting from 15,000 ft. is 438 lbs.[19] The mass required increases enormously with decreasing efficiency, for, with but half of the former " effective velocity " (3,500 ft./sec.) the " total initial mass," even for a start from 15,000 ft., is 351,000 lbs. The masses would obviously be slightly less if the acceleration exceeded 150 ft./sec.$^2$

It is of interest to speculate upon the possibility of proving that such extreme altitudes had been reached even if they actually were attained. In general, the proving would be a difficult matter. Thus, even if a mass of flash powder, arranged to be ignited automatically after a long interval of time, were projected vertically upward, the light would at best be very faint, and it would be difficult to foretell, even approximately, the direction in which it would be most likely to appear.

The only reliable procedure would be to send the smallest mass of flash powder possible to the dark surface of the moon when in conjunction (*i. e.*, the " new " moon), in such a way that it would be ignited on impact. The light would then be visible in a powerful telescope. Further, the larger the aperture of the telescope, the greater would be the ease of seeing the flash, from the fact that a telescope enhances the brightness of point sources, and dims a faint background.

An experiment was performed to find the minimum mass of flash powder that should be visible at any particular distance. In order to reproduce, approximately, the conditions that would obtain at the surface of the moon, the flash powder was placed in small capsules, C, plate 9, figure 1, held in glass tubes, T, closed by rubber stoppers. The tubes were exhausted to a pressure of from 3 to 10 cm. of mercury, and sealed, the stoppers being painted with wax, to preserve the vacuum. Two shellacked wires, passing to the powder, permitted firing of the powder by an automobile spark coil.

It was found that Victor flash powder was slightly superior to a mixture of powdered magnesium and sodium nitrate, in atomic proportions, and much superior to a mixture of powdered magnesium and potassium chlorate, also in atomic proportions.

In the actual test, six samples of Victor flash powder, varying in weight from 0.05 gram to 0.0029 gram were placed in tubes as shown in plate 9, figure 1, and these tubes were fastened in blackened compartments of a box, plate 9, figure 2, and plate 10, figure 1. The ignition system was placed in the back of the same box, as shown in plate 10, figure 2. This system comprised a spark coil, operated by three triple cells of " Ever-ready " battery, placed two by two in parallel. The charge was fired on closing the primary switch at the left. The six-point switch at the right served to connect the tubes, in order, to the high-tension side of the coil.

The flashes were observed at a distance of 2.24 miles on a fairly clear night; and it was found that a mass of 0.0029 grams of Victor flash powder was visible, and that 0.015 gram was strikingly visible, all the observations being made with the unaided eye. The minimum mass of flash powder visible at this distance is thus surprisingly small.

From these experiments it is seen that if this flash powder were exploded on the surface of the moon, distant 220,000 miles, and a telescope of one foot aperture were used—the exit pupil being not greater than the pupil of the eye (e. g., 2 mm.)—we should need a mass of flash powder of

2.67 lbs., to be just visible, and
13.82 lbs. or less, to be strikingly visible.

If we consider the final mass of the last "secondary" rocket plus the mass of the flash powder and its container, to be four times the mass of the flash powder alone, we should have, for the *final mass of the rocket,* four times the above masses. These final masses correspond to the "one pound final mass" which has been mentioned throughout the calculations.

The "total initial masses," or the masses necessary for the start at the earth, are at once obtained from the data given in table VII. Thus if the start is made from sea-level, and the "effective velocity of ejection" is 7,000 ft./sec., we need 602 lbs. for every pound that is to be sent to "infinity."[1]

We arrive, then, at the conclusion that the "total initial masses" necessary would be

6,436 lbs. or 3.21 tons; flash just visible, and

33,278 lbs. or 16.63 tons (or less); flash strikingly visible.

A "total initial mass" of 8 or 10 tons would, without doubt, raise sufficient flash powder for clear visibility.[21]

These masses could, of course, be much reduced by the employment of a larger telescope. For example, with an aperture of two feet, the masses would be reduced to one-fourth of those just given. The use of such a large telescope would, however, limit considerably the possible number of observers. In all cases, the magnification should be so low that the entire lunar disk is in the field of the telescope.

It should be added that the probability of collision of a small object with meteors of the visible type is negligible, as is indicated in Appendix G, page 64.

This plan of sending a mass of flash powder to the surface of the moon, although a matter of much general interest, is not of obvious scientific importance. There are, however, *developments of the general method under discussion, which involve a number of important features not herein mentioned,* which could lead to results of much scientific interest. These developments involve many experimental difficulties, to be sure; but they depend upon nothing that is really impossible.

---

[1] A simple calculation [20] will show that the total initial mass required to send one pound to the surface of the moon is but slightly less than that required to send the mass to "infinity."

## SUMMARY

1. An important part of the atmosphere, that extends for many miles beyond the reach of sounding balloons, has up to the present time been considered inaccessible. Data of great value in meteorology and in solar physics could be obtained by recording instruments sent into this region.

2. The rocket, in principle, is ideally suited for reaching high altitudes, in that it carries apparatus without jar, and does not depend upon the presence of air for propulsion. A new form of rocket apparatus, which embodies a number of improvements over the common form, is described in the present paper.

3. A theoretical treatment of the rocket principle shows that, if the velocity of expulsion of the gases were considerably increased and the ratio of propellant material to the entire rocket were also increased, a tremendous increase in range would result, from the fact that these two quantities enter exponentially in the expression for the initial mass of the rocket necessary to raise a given mass to a given height.

4. Experiments with ordinary rockets show that the efficiency of such rockets is of the order of 2 per cent, and the velocity of ejection of the gases, 1,000 ft./sec. For small rockets the values are slightly less.

With a special type of steel chamber and nozzle, an efficiency has been obtained with smokeless powder of over 64 per cent (higher than that of any heat engine ever before tested) ; and a velocity of nearly 8,000 ft./sec., which is the highest velocity so far obtained in any way except in electrical discharge work.

5. Experiments were repeated with the same chambers *in vacuo,* which demonstrated that the high velocity of the ejected gases was a real velocity and not merely an effect of reaction against the air. In fact, experiments performed at pressures such as probably exist at an altitude of 30 miles gave velocities even higher than those obtained in air at atmospheric pressure, the increase in velocity probably being due to a difference in ignition. Results of the experiments indicate also that this velocity could be exceeded, with a modified form of apparatus.

6. Experiments with a large chamber demonstrated that not only are large chambers operative, but that the velocities and efficiencies are higher than for small chambers.

7. A calculation based upon the theory, involving data that is in part that obtained by experiments, and in part what is considered as realizable in practice, indicates that the initial mass required to raise recording instruments of the order of one pound, even to the extreme upper atmosphere, is moderate. The initial mass necessary is likewise not excessive, even if the effective velocity is reduced by half. Calculations show, however, that any apparatus in which ordinary rockets are used would be impracticable owing to the very large initial masses that would be required.

8. The recovery of the apparatus, on its return, need not be a difficult matter, from the fact that the time of ascent even to great altitudes in the atmosphere will be comparatively short, due to the high speed of the rocket throughout the greater part of its course. The time of descent will also be short; but free fall can be satisfactorily prevented by a suitable parachute. A parachute will be operative for the reason that high velocities and small atmospheric densities are essentially the same as low velocities and ordinary density.

9. Even if a mass of the order of a pound were propelled by the apparatus under consideration until it possessed sufficient velocity to escape the earth's attraction, the initial mass need not be unreasonably large, for an effective velocity of ejection which is without doubt attainable. A method is suggested whereby the passage of a body to such an extreme altitude could be demonstrated.

## CONCLUSION

Although the present paper is not the description of a working model, it is believed, nevertheless, that the theory and experiments, herein described, together settle all points that could seriously be questioned, and that it remains only to perform certain necessary preliminary experiments before an apparatus can be constructed that will carry recording instruments to any desired altitude."

# APPENDIX A

## THEORY OF THE MOTION WITH DIRECT LIFT

Let $M$ = the mass of the suspended system, comprising the chamber together with any parts rigidly attached thereto,

$m_0$ = the mass of the expelled charge, comprising wadding and the attached copper wire, the smokeless powder charge (and also, in the experiments *in vacuo,* the black powder priming charge),

$V$ = the initial upward velocity of the mass $M$,

$v$ = the *average* downward velocity of the mass $m_0$

and $s$ = the upward displacement of the mass $M$.

We have at once for the initial velocity of the mass $M$,

$$V^2 = 2gs,$$

and employing the Conservation of Momentum, we have for the kinetic energy per gram of mass $m_0$, expelled,

$$\frac{v^2}{2} = \frac{M^2}{m_0^2} gs.$$

# APPENDIX B

## THEORY OF THE DISPLACEMENTS FOR SIMPLE HARMONIC MOTION

In addition to the notation given under Appendix A, the following additional notation must be employed:

Let $m_s$ = the mass of the spring,

$F_1$ = the force in dynes which produces unit extension of the spring,

$m_1$ = the mass in dynes which produces unit extension of the spring,

and $s$ = the upward displacement of $M$, resulting from the firing, that would be had if there were no friction.

Then, allowing for the mass of the spring, we have, from the theory of simple harmonic motion:

$$Fx = \left(M + \frac{m_s}{3}\right)\left(\frac{2\pi}{P}\right)^2 x,$$

where $x$ is any displacement, and $P$ is the period of the motion.

But V is the maximum velocity during the motion and hence $V = \omega.s$, where s is the maximum displacement, and $\omega$ is a constant, having the usual significance; also

$$P = \frac{2\pi}{\omega}.$$

Hence

$$m_1 g = \left(M + \frac{m_s}{3}\right) \frac{V^2}{s^2}.$$

But by the Conservation of Linear Momentum,

$$\left(M + \frac{m_s}{3}\right) V = m_0 v.$$

Hence

$$m_1 g = \left(M + \frac{m_s}{3}\right) \left(\frac{m_0 v}{M + \frac{m_s}{3}}\right)^2 \frac{1}{s^2},$$

giving, for the kinetic energy per gram of mass expelled,

$$\frac{v^2}{2} = \frac{\left(M + \frac{m_s}{3}\right)(m_1 g)}{2 m_0^2} s^2.$$

From this it is possible to obtain the efficiency, by dividing by the heat value of the powder, in ergs; and also the velocity in kilometers per second by multiplying by two, extracting the square root, and dividing by $10^5$.

## CORRECTION OF THE DISPLACEMENT, s, FOR FRICTION

The displacement, s, in the preceding calculation is assumed to be the corrected displacement. This is obtained from the upward displacement $s_1$, and the downward displacement $s_2$, as

$$s = s_1 \sqrt{\frac{s_1}{s_2}}$$

## APPENDIX C
### THEORY OF DIRECT-LIFT IMPULSE-METER

The theory of the direct-lift impulse-meter is as follows:

Calling I, the momentum of the gas that strikes the end of the aluminium cylinder,

$m_c =$ the mass of the aluminium cylinder,

$V_c =$ the initial upward velocity of the cylinder,

$A_a =$ the area of cross-section of the cylinder,

$A_g =$ the maximum area of cross-section of the suspended system comprising the gun, lead weight, and holders,

and $s =$ the displacement of the aluminium cylinder, as obtained from the trace on the smoked glass tube, we have, by the principle of the Conservation of Linear Momentum, for the momentum per unit area produced by the gaseous rebound,

$$\frac{I}{A_c} = \frac{m_c V_c}{A_c} = \frac{m_c \sqrt{2gs}}{A_c}$$

Hence the momentum communicated to the suspended system by the gaseous rebound is

$$\frac{m_c A_g \sqrt{2gs}}{A_c},$$

and calling Q the ratio of the momentum given the gun by gaseous rebound to the observed momentum of the suspended system, we have

$$Q = \frac{m_c A_g \sqrt{2gs}}{m_0 A_c v}.$$

## APPENDIX D

### THEORY OF SPRING IMPULSE-METER

The theory of the spring impulse-meter is as follows: If we use the same notation as in the preceding case, calling, in addition, the mass of the spring $m_s$, and the mass required for unit extension of the spring, $m_1$, we have, by the same theory as that for the gun suspended by a spring,

$$V_c = \frac{\sqrt{m_1 g}}{\sqrt{m_c + \frac{1}{3} m_s}} s.$$

Hence the momentum per unit area, communicated to the upper cap of the 12-inch pipe, when the chamber is fired, is

$$\frac{I}{A_c} = \frac{(m_c + \frac{1}{3} m_s) V_c}{A_c} = \frac{\sqrt{m_c + \frac{1}{3} m_s} \sqrt{m_1 g s}}{A_c}$$

Hence the momentum that would be communicated to the *suspended system* by the gaseous rebound, *provided the system were at the top of the 12-inch pipe,* would be

$$\frac{A_g \sqrt{m_c + \frac{1}{3} m_s} \sqrt{m_1 g s}}{A_c},$$

and the percentage, Q, of the momentum communicated by the gaseous rebound to the observed momentum of the suspended system, is

$$Q = \frac{A_g \sqrt{m_c + \frac{1}{3} m_s} \sqrt{m_1 g}}{A_c m_0 v} s.$$

## APPENDIX E

### CHECK ON APPROXIMATE METHOD OF CALCULATION, FOR SMALL CHARGES FIRED IN RAPID SUCCESSION

Consider a rocket weighing 10 lbs., having 2 lbs. of propelling material, fired two ounces at a time, eight times per second, with a velocity of 6,000 ft./sec.—much less than the highest velocity attained in the experiments, either in air or *in vacuo*.

Let us suppose that, for simplicity, the rocket is directed upward and that each shot takes place instantly (a supposition not far from the truth) ; the velocity remaining constant between successive shots.

After the first shot, the mass, $9\frac{7}{8}$ lbs., has an upward velocity $v_0$ due to the downward velocity of the $\frac{1}{8}$ lb. expelled. This velocity, $v_0$, is at once found by the Conservation of Momentum. But it is decreased by gravity until, at the end of $\frac{1}{8}$ sec., it is reduced to

$$v_0' = v_0 - gt,$$

the space passed over during this time being

$$s = v_0 t - \tfrac{1}{2}gt^2.$$

We have then, $v_0' = 71.8$ ft./sec., and $s = 9.23$ ft.

At the beginning of the second interval of $\frac{1}{8}$ sec., an additional velocity is given the remaining mass, of 76.8 ft./sec., and the final velocity and space passed over may be found in the same way. By completing the calculations for the remaining intervals we shall have
for time just under $\frac{1}{2}$ sec., $v' = 293.1$ ft./sec.; s= 91.98 ft.
for time just under 1 sec., $v_0' = 603.8$ ft./sec.; s= 335.48 ft.
and
for time just under 2 sec., $v_0' = 1284.1$ ft./sec.; s=1315.68 ft./sec.

These figures compare well with those in table V, for $s_1$. In the present check, air resistance would doubtless be unimportant until the velocity had reached 1,000 ft./sec. or so ; but the velocity would, even if decreased somewhat by air resistance, compare favorably with that of a projectile fired from a gun.

No more elaborate calculation is necessary to demonstrate the importance of the device, even for military purposes alone ; for it combines portability and cheapness (no gun is required for firing it) with a range which compares favorably with the best artillery. Further, all difficulties of the nature of erosion are, of course, avoided.

## APPENDIX F

### PROOF THAT THE RETARDATION BETWEEN 500,000 FT. AND 1,000,000 FT. IS NEGLIGIBLE

The falling-off of velocity, w, due to air resistance, is given by

$$P \frac{\rho}{\rho_0} sh = \tfrac{1}{2} M_0 w_1^2$$

where $P$ = the mean air resistance in poundals per square inch between the altitudes 500,000 and 1,000,000 ft. from the previously mentioned velocity curves, the pressure being considered as atmospheric.

$\rho$ = the mean density over this distance,

$s$ = the mean area of cross-section of the apparatus throughout the distance, taken as 25 square inches in view of the average mass, $M_0$, throughout the interval, and

$h$ = the distance traversed: 500,000 ft.

It is thus found that the loss of velocity w is less than 10 ft./sec. (for a = 150 ft./sec.) even when $\frac{\rho}{\rho_0}$ is taken as constant throughout the distance and equal to that at 500,000 ft. (*i. e.*, $2.73 \times 10^{-9}$).

## APPENDIX G

### PROBABILITY OF COLLISION WITH METEORS

The probability of collision with meteors of " visible " size is negligible. This can be shown by deriving an expression for the probability of collision of a sphere with particles moving in directions at random, all having constant velocity, the expression being obtained on the assumption that the speed of the sphere is small compared with the speed of the particles.

If we accept Newton's estimate [1] of the average distance apart of meteors as being 250 miles, we have by considering collision between very small meteors of velocity 30 miles/sec., and a sphere one foot in diameter of velocity one mile/sec., moving over a distance of 220,000 miles, the probability [23] as $1.23 \times 10^{-8}$; which is, of course, practically negligible. The value would be slightly greater if the meteors were considered as having a diameter of several centimeters, rather than being particles [24]; but the probability would be less, however, if meteor swarms were avoided.

---

[1] Newton, Encyc. Brit. 9 ed. v. 16.

# NOTES

[10] A step-by-step method of solution similar to that herein employed can evidently be used for oblique projection—other conditions remaining the same.

[11] If the efficiency is estimated by the kinetic energy of the rocket itself (from the velocity the average mass of the rocket would acquire, by virtue of the recoil of the gases ejected with the "average velocity" measured), the efficiencies will, of course, be less than the two values given in table I, being, respectively, 0.39 and 0.50 per cent.

[12] Since this manuscript was written, rockets with a single charge, constructed along the general lines here explained, have been considerably further developed.

[13] Chambers of considerably reduced weight have since been made and tested for velocities comparable to those here mentioned. For two particular types of loading device, the ratio of weight of chamber to weight of charge (here, 120) were, respectively, 63 (also 30 for this case, but at a sacrifice of velocity) and 22; the ratio, for the nozzles, being reducible to comparatively small values. In neither of these cases was any special attempt made to reduce the weight of the chambers.

[14] Later experiments support this prediction, and also demonstrate that firing of the charges can take place in rapid succession.

[15] The values of c and $(1-k)$, here assigned, were chosen as being the largest that could reasonably be expected. Later experiments have shown that lower values are more easily realizable, but it should at the same time be understood that *no special attempt has been made to obtain experimentally the highest values of these quantities.* The numbers chosen may, then, be considered as at least possible limiting values.

It is well to mention, in this connection, that the developments with the multiple-charge rocket have, so far, exceeded original expectations. This is in accord with the fact that the experimental results have, from the start, been more favorable than were expected. Thus an efficiency of 50 per cent was at first considered the limit of what could be attained, and 4,000 to 5,000 ft./sec., the highest possible velocity. Further it was naturally not expected that the velocities obtained *in vacuo* would actually exceed those in air; nor were chambers as light as those at present used considered producible without considerable experimental difficulty.

[16] *Distribution of mass among the secondary rockets for cases of large total initial mass.*—For very great altitudes, secondary rockets will be necessary, as already explained, in order to keep the proportion of propellant to total weight sensibly constant. The most extreme cases will require groups of secondary rockets, which groups are discharged in succession.

There are, under any circumstances, two possibilities: either the secondaries may be small, so that each time a secondary rocket, or group of secondaries, is discarded, the total mass is not appreciably changed, as indicated schematically at (a), figure 8; or a series of as large secondaries as possible may be used, (b), figure 8, in which case the empty casings constitute a considerable fraction of the entire weight at the time the discarding takes place.

In so far as avoiding difficulties of construction are concerned, the use of a smaller number of larger secondaries is preferable, but they should be long and narrow, as otherwise the air resistance on the nearly empty casings will be greater for the same weight of propellant than would be the case if groups of small secondaries, case (a), were used, in as compact an arrangement as

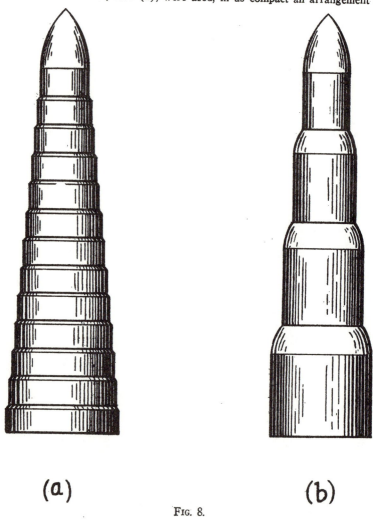

(a)            (b)

Fig. 8.

possible. It should be explained, also, that if very small secondaries were employed, the metal of the magazines and casings would become a considerable fraction of the entire weight, as the amount of surface enclosing the propellant would then be a maximum.

*Possibility of employing case (b).*—A rough calculation shows at once the possibility of using a comparatively small number of large secondaries (or groups), *provided,* as is, of course, to be expected from dimensional considerations, that the larger any individual rocket, the less, in proportion, need be the ratio of weight of metal to weight of propellant.

Such a calculation can be made by finding the number of secondary rockets, for case (b), that would be required for the same total initial mass, other conditions being the same, as for continuous loss of mass with zero relative velocity, which is practically case (a).

For the latter, equation (7), in which R and g are neglected, is evidently sufficient for the purpose, for the reason that the form of the expression, so far as (1-k) is concerned, is the same whether or not R and g is included.

Let us now find what conditions must hold for case (b), in order that the total initial mass shall equal that for case (a). Assume, first, that the casings are discarded successively at the end of n equal intervals of time, no mass being discarded except at these times; the velocity of gas ejection being c, as before. The total initial mass is obtained as the product of the initial masses for each interval, from equation (7) with k = 0, assuming the final mass for each interval is, as before, 1 lb., after first multiplying the initial masses by a greater factor than unity, the excess over unity being the weight h, of the casings which are discarded at the end of the intervals.

If, in case (a), we divide the time into n equal intervals in the same way, we shall have, as the condition that the total initial masses are the same in the two cases,

$$M = e^{\frac{a(t/n)n}{c(1-k)}} = (1 + h)^n e^{\frac{a(t/n)n}{c}} \qquad (15)$$

We obtain, then, on combining (15) with (7),

$$M^k = (1 + h)^n,$$

from which

$$n = k \frac{\log M}{\log (1+h)} \qquad (16)$$

Let us assume, for case (a) (many small secondary rockets), as well as for case (b) (large secondary rockets), that the ratio of mass of metal to mass of propellant is the minimum reasonable amount that can be expected, which may be put tentatively, at least, as 1/14 and 1/18, respectively.

Two cases will suffice for purpose of illustration: one in which the ratio of initial to final mass is moderately large, *e. g.,* 40, and the other in which the ratio is extreme, *e. g.,* 600.

The numbers of secondaries (or separate groups) for (b), for these two cases, are, from (16), 5 and 9 respectively, n being necessarily an integer.

It is to be understood that the numbers could be made even smaller, although this would necessitate larger total initial masses.

[17] If the start were made at a greater elevation than 15,000 feet, for example, at 20,000 or 25,000 feet, the reduction of the "total initial mass" would, of course, be considerably greater. Further, if the rocket were of comparatively small mass, it could be raised to an even greater initial height by balloons.

[18] Actually, 300 grams would be sufficient, for many researches.

[19] Attention is called to the fact that hydrogen and oxygen, combining in atomic proportions, afford the greatest heat per unit mass of all chemical transformations. For this reason, if the calculations are made under the

assumption that hydrogen and oxygen are used (in the liquid or solid state, to avoid weight of the container), giving the same efficiency as that for which "Infallible" smokeless powder produces respective velocities of, for example, 5,500 ft./sec. and 7,500 ft./sec., the velocities (deducting 218.47 calories per gram as the latent heat plus specific heat, from boiling point to ordinary temperature) would be 9,400 ft./sec. and 11,900 ft./sec.; and the total initial masses for a start from 15,000 feet, respectively, 119 pounds and 43.5 pounds.

Incidentally, except for difficulties of application, the use of hydrogen and oxygen would have several other evident advantages.

[20] This calculation is made under the assumption of stationary centers for the earth and moon.

[21] The time of transit for the case under discussion would, of course, be comparatively large. If, however, the final velocity were to exceed by 1,000 or 2,000 ft./sec. the velocity calculated, the time would be reduced to a day or two.

The time can be calculated from the solution, by Plana (Memoire della Reale Accademia della Scienze di Torino, Ser. 2, vol. 20, 1863, pp. 1-86), of the analogous problem of the determination of the initial velocity and time of transit of a body, such as a volcanic rock, projected from the moon toward the earth.

[22] At the time of signing of the armistice, the net result of the development of a reloading mechanism had been the demonstration of an operative apparatus that was simple and travelled straight, with the essential parts sufficiently strong and light, using a few cartridges of simple form.

The work remaining, upon which progress has since been made, has been the adaption of the device for a large proportionate weight of propellant.

[23] The probable number of collisions here calculated is the sum of the probable numbers obtained by taking the velocity of the spherical body, and of the meteors, separately equal to zero.

Let $v$ = velocity of the spherical body,

$V$ = velocity of the meteors,

$n$ = the number of meteors per unit volume, which number is, of course, a fraction (mutual collisions between meteors being neglected), and

$S$ = the area of cross-section of the spherical body.

For $v = 0$, the meteors, if any, which strike the sphere during the time $t$ to $t + dt$ will have come from a spherical shell of radii $Vt$ and $V(t + dt)$, neglecting the diameter of the spherical body in comparison with that of the spherical shell. Further, the probable number in any small volume, in this shell, which are so directed as to strike the body, is

$$\frac{S}{4\pi V^2 t^2} \; ;$$

being the ratio of the solid angle subtended at the element, by the spherical body, to the whole solid angle, $4\pi$. Hence the probable number of collisions, $N$, from all directions, between the time $t_1$ and $t_2$ is, evidently,

$$N = nSV(t_2 - t_1).$$

For $V = 0$, an expression of the same form is obtained for the probable number of meteors within the space swept out by the spherical body.

The sum of these separate probable numbers is the number 1.23 × 10⁻⁸, in Appendix G.

In general, for any values of v and V, the meteors reaching the spherical body at successive instants come from a spherical surface of increasing radius, Vt, with moving center distant vt in front of the initial position of the spherical body.

It should be explained that when v differs but little from V, the relative velocity of the body and meteors is small enough to be neglected, for meteors on this expanding spherical surface lying outside a certain cone, the vertex of which coincides with the moving center of the spherical body.

Also, if v exceeds V, the only part of the expanding spherical surface which is to be considered is that lying outside the contact circle of the tangent cone, the vertex of which also coincides with the moving center of the spherical body.

Attention is called to the fact that, even if meteor swarms were not avoided, the probable number of collisions would be reduced if the direction of motion were substantially that of the swarm.

²⁴ No difference in the calculation would be necessary if the radius of the sphere were to be increased by the diameter of the meteors, these being then considered as particles.

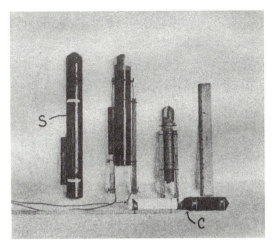

FIG. 1

FIG. 2

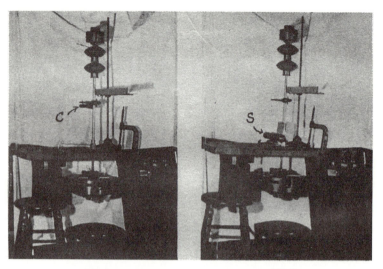

FIG. 1 FIG. 2

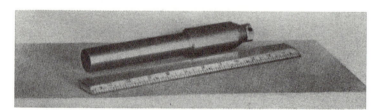

FIG. 3

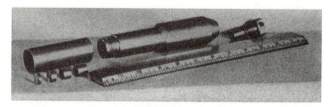

FIG. 4

FIG. 5

FIG. 1

FIG. 2

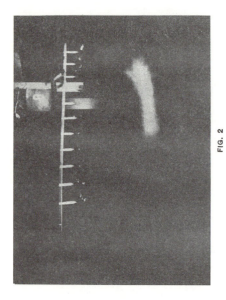

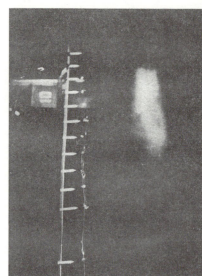

FIG. 2

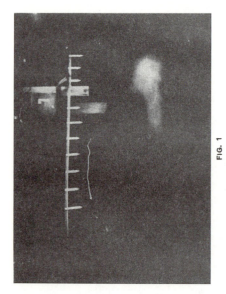

FIG. 1

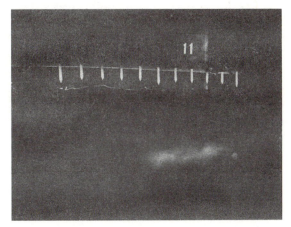

FIG. 1

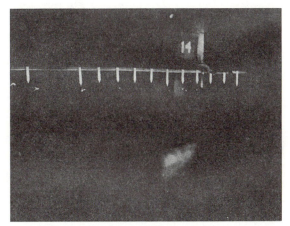

FIG. 2

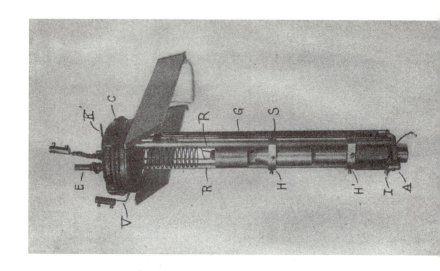

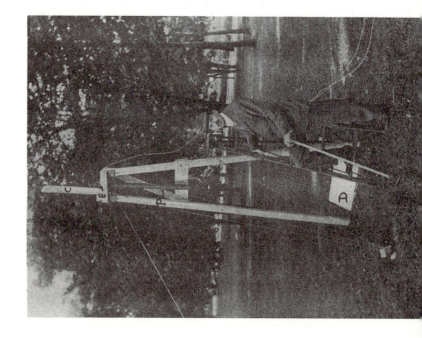

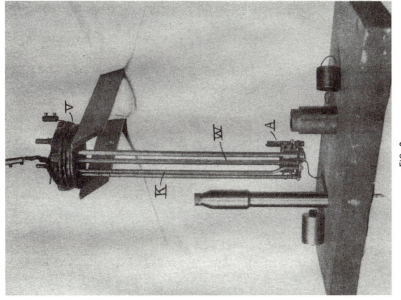

FIG. 2

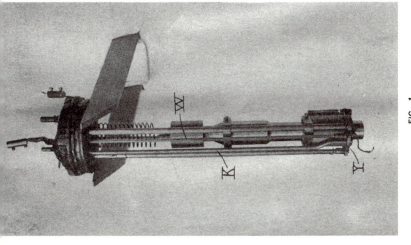

FIG. 1

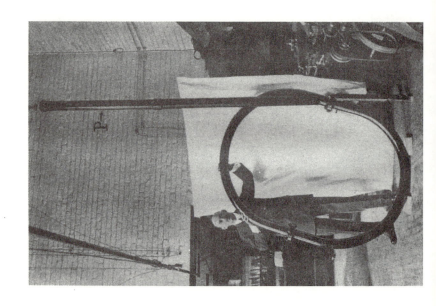

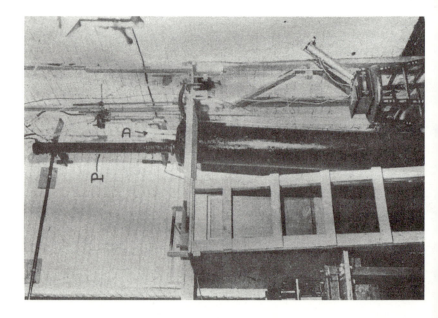

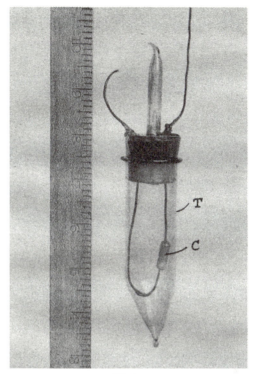

FIG. 1

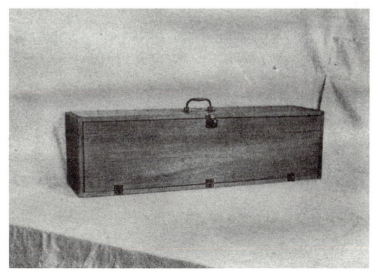

FIG. 2

FIG. 1

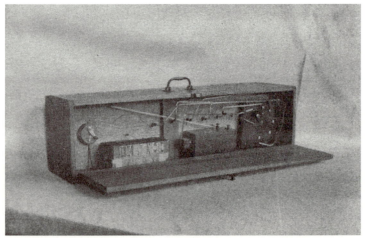

FIG. 2

# LIQUID-PROPELLANT ROCKET
# DEVELOPMENT

LIQUID-PROPELLANT ROCKET DEVELOPMENT was written late in 1935 as a report to the Daniel and Florence Guggenheim Foundation on the work accomplished toward the goal of high-altitude liquid-fuel rockets. Between his 1919 paper and this report Dr. Goddard issued no formal public statement regarding details or the progress of his research. Beginning about September 1930 his work was carried on principally at Mescalero Ranch, Roswell, New Mexico, with the support of grants from the Carnegie Institution of Washington, the late Daniel Guggenheim, and the Daniel and Florence Guggenheim Foundation. In this report the physicist disclosed for the first time the beginnings of his work with liquid propellants, gave details of the world's first liquid-fuel rocket, which he shot on March 16, 1926 at Auburn, Mass., after experiments with liquid fuels which began in 1920, and revealed the astounding progress he had been able to make, on what now seem relatively slender resources, toward the development of high altitude rockets.

*Reproduced here by facsimile printing exactly as originally published by the Smithsonian Institution in 1936*

SMITHSONIAN MISCELLANEOUS COLLECTIONS

VOLUME 95, NUMBER 3

# LIQUID-PROPELLANT ROCKET DEVELOPMENT

(WITH 11 PLATES)

BY

ROBERT H. GODDARD

Clark University

(PUBLICATION 3381)

CITY OF WASHINGTON

PUBLISHED BY THE SMITHSONIAN INSTITUTION

MARCH 16, 1936

# LIQUID-PROPELLANT ROCKET DEVELOPMENT

By ROBERT H. GODDARD

*Clark University*

(WITH 11 PLATES)

The following is a report made by the writer to the Daniel and Florence Guggenheim Foundation concerning the rocket development carried out under his direction in Roswell, N. Mex., from July 1930 to July 1932, and from September 1934 to September 1935, supported by this Foundation.

This report is a presentation of the general plan of attack on the problem of developing a sounding rocket, and of the results obtained. Further details will be set forth in a later paper, after the main objects of the research have been attained.

## INTRODUCTION

In a previous paper[1] the author developed a theory of rocket performance and made calculations regarding the heights that might reasonably be expected for a rocket having a high velocity of the ejected gases and a mass at all times small in proportion to the weight of propellant material. It was shown that these conditions would be satisfied by having a tapered nozzle through which the gaseous products of combustion were discharged,[2] by feeding successive portions of propellant material into the rocket combustion chambers,[3] and further by employing a series of rockets, of decreasing size, each fired when the rocket immediately below was empty of fuel.[3] Experimental results with powder rockets were also presented in this paper.

Since the above was published, work has been carried on for the purpose of making practical a plan of rocket propulsion set forth in 1914[3] which may be called the liquid-propellant type of rocket. In this rocket, a liquid fuel and a combustion-supporting liquid are fed under pressure into a combustion chamber provided with a conical nozzle through which the products of combustion are discharged.

---

[1] Smithsonian Misc. Coll., vol. 71, no. 2, 1919.

[2] U. S. Patent, Rocket Apparatus, No. 1,102,653, July 7, 1914.

[3] U. S. Patent, Rocket Apparatus, No. 1,103,503, July 14, 1914.

The advantages of the liquid-propellant rocket are that the propellant materials possess several times the energy of powders, per unit mass, and that moderate pressures may be employed, thus avoiding the weight of the strong combustion chambers that would be necessary if propulsion took place by successive explosions.

Experiments with liquid oxygen and various liquid hydrocarbons, including gasoline and liquid propane, as well as ether, were made during the writer's spare time from 1920 to 1922, under a grant by Clark University. Although oxygen and hydrogen, as earlier suggested, possess the greatest heat energy per unit mass, it seems likely that liquid oxygen and liquid methane would afford the greatest heat value of the combinations which could be used without considerable difficulty. The most practical combination, however, appears to be liquid oxygen and gasoline.

In these experiments it was shown that a rocket chamber and nozzle, since termed a "rocket motor," could use liquid oxygen together with a liquid fuel, and could exert a lifting force without danger of explosion and without damage to the chamber and nozzle. These rockets were held by springs in a testing frame, and the liquids were forced into the chamber by the pressure of a noninflammable gas.

The experiments were continued from 1922 to 1930, chiefly under grants from the Smithsonian Institution. Although this work will be made the subject of a later report, it is desirable in the present paper to call attention to some of the results obtained. On November 1, 1923, a rocket motor operated in the testing frame, using liquid oxygen and gasoline, both supplied by pumps on the rocket.

In December 1925 the simpler plan previously employed of having the liquids fed to the chamber under the pressure of an inert gas in a tank on the rocket was again employed, and the rocket developed by means of these tests was constructed so that it could be operated independently of the testing frame.

The first flight of a liquid oxygen-gasoline rocket was obtained on March 16, 1926, in Auburn, Mass., and was reported to the Smithsonian Institution May 5, 1926. This rocket is shown in the frame from which it was fired, in plate 1, figure 1. Pressure was produced initially by an outside pressure tank, and after launching by an alcohol heater on the rocket.

It will be seen from the photograph that the combustion chamber and nozzle were located forward of the remainder of the rocket, to which connection was made by two pipes. This plan was of advantage

---

[4] Smithsonian Misc. Coll., vol. 71, no. 2, 1919.

in keeping the flame away from the tanks, but was of no value in producing stabilization. This is evident from the fact that the direction of the propelling force lay along the axis of the rocket, and not in the direction in which it was intended the rocket should travel, the condition therefore being the same as that in which the chamber is at the rear of the rocket. The case is altogether different from pulling an object upward by a force which is constantly vertical, when stability depends merely on having the force applied above the center of gravity.

Plate 1, figure 2 shows an assistant igniting the rocket, and plate 2, figure 1 shows the group that witnessed the flight, except for the camera operator. The rocket traveled a distance of 184 feet in 2.5 seconds, as timed by a stop watch, making the speed along the trajectory about 60 miles per hour.

Other short flights of liquid oxygen-gasoline rockets were made in Auburn, that of July 17, 1929, happening to attract public attention owing to a report from someone who witnessed the flight from a distance and mistook the rocket for a flaming airplane. In this flight the rocket carried a small barometer and a camera, both of which were retrieved intact after the flight (pl. 2, fig. 2). The combustion chamber was located at the rear of the rocket, which is, incidentally, the best location, inasmuch as no part of the rocket is in the high velocity stream of ejected gases, and none of the gases are directed at an angle with the rocket axis.

During the college year 1929-30 tests were carried on at Fort Devens, Mass., on a location which was kindly placed at the disposal of the writer by the War Department. Progress was made, however, with difficulty, chiefly owing to transportation conditions in the winter. ·

At about this time Col. Charles A. Lindbergh became interested in the work and brought the matter to the attention of the late Daniel Guggenheim. The latter made a grant which permitted the research to be continued under ideal conditions, namely, in eastern New Mexico; and Clark University at the same time granted the writer leave of absence. An additional grant was made by the Carnegie Institution of Washington to help in getting established.

It was decided that the development should be carried on for 2 years, at the end of which time a grant making possible 2 further years' work would be made if an advisory committee, formed at the time the grant was made, should decide that this was justified by the results obtained during the first 2 years. This advisory committee was as follows: Dr. John C. Merriam, chairman; Dr. C. G. Abbot; Dr. Walter S. Adams; Dr. Wallace W. Atwood; Col. Henry Breckinridge; Dr. John A. Fleming; Col. Charles A. Lindbergh; Dr. C. F. Marvin; and Dr. Robert A. Millikan.

## THE ESTABLISHMENT IN NEW MEXICO

Although much of the eastern part of New Mexico appeared to be suitable country for flights because of clear air, few storms, moderate winds, and level terrain, it was decided to locate in Roswell, where power and transportation facilities were available.

A shop 30 by 55 feet was erected in September 1930 (pl. 3, figs. 1, 2), and the 60-foot tower previously used in Auburn and Fort Devens was erected about 15 miles away (pl. 4, fig. 1). A second tower, 20 feet high (pl. 4, fig. 2), was built near the shop for static tests, that is, those in which the rocket was prevented from rising by heavy weights, so that the lift and general performance could be studied. These static tests may be thought of as " idling " the rocket motor. A cement gas deflector was constructed under each tower, as may be seen in plate 4, figures 1, 2, whereby the gases from the rocket were directed toward the rear, thus avoiding a cloud of dust which might otherwise hide the rocket during a test.

## STATIC TESTS OF 1930–32

Although, as has been stated, combustion chambers had been constructed at Clark University which operated satisfactorily, it appeared desirable to conduct a series of thorough tests in which the operating conditions were varied, the lift being recorded as a function of the time. Various modifications in the manner of feeding the liquids under pressure to the combustion chamber were tested, as well as variations in the proportions of the liquids, and in the size and shape of the chambers. The chief conclusions reached were that satisfactory operation of the combustion chambers could be obtained with considerable variation of conditions, and that larger chambers afforded better operation than those of smaller size.

As will be seen from plate 4, figure 2, the supporting frame for the rocket was held down by four steel barrels containing water. Either two or four barrels could be filled, and in the latter case the total weight was about 2,000 lbs. This weight was supported by a strong compression spring, which made possible the recording of the lift on a revolving drum (pl. 5, fig. 1) driven by clockwork.

The combustion chamber finally decided upon for use in flights was 5¾ inches in diameter and weighed 5 pounds. The maximum lift obtained was 289 pounds, and the period of combustion usually exceeded 20 seconds. The lifting force was found to be very steady, the variation of lift being within 5 percent.

The masses of liquids used during the lifting period were the quantities most difficult to determine. Using the largest likely value of the total mass of liquids ejected and the integral of the lift-time curve obtained mechanically, the velocity of the ejected gases was estimated to be over 5,000 feet per second. This gave for the mechanical horsepower of the jet 1,030 hp., and the horsepower per pound of the combustion chamber, considered as a rocket motor, 206 hp. It was found possible to use the chambers repeatedly.

The results of this part of the development were very important, for a rocket to reach great heights can obviously not be made unless a combustion chamber, or rocket motor, can be constructed that is both extremely light and can be used without danger of burning through or exploding.

## FLIGHTS DURING THE PERIOD 1930–32

The first flight obtained during this period was on December 30, 1930, with a rocket 11 feet long, weighing 33.5 pounds. The height obtained was 2,000 feet, and the maximum speed was about 500 miles per hour. A gas pressure tank was used on the rocket to force the liquid oxygen and the gasoline into the combustion chamber.

In further flights pressure was obtained by gas pressure on the rocket, and also by pumping liquid nitrogen through a vaporizer, the latter means first being employed in a flight on April 19, 1932.

In order to avoid accident, a remote control system was constructed in September 1931, whereby the operator and observers could be stationed 1,000 feet from the tower, and the rocket fired and released at will from this point. This arrangement has proved very satisfactory. Plate 5, figure 2 shows the cable being unwound between the tower and the 1,000-foot shelter, the latter being seen in the distance, and plate 6, figure 1 shows the control keys being operated at the shelter, which is provided with sand bags on the roof as protection against possible accident. Plate 5, figure 2 shows also the level and open nature of the country.

One observer was stationed 3,000 feet from the tower, in the rear of the 1,000-foot shelter, with a recording telescope (pl. 6, fig. 2). Two pencils attached to this telescope gave a record of the altitude and azimuth, respectively, of the rocket, the records being made on a paper strip, moved at a constant speed by clockwork. The sights at the front and rear of the telescope, similar to those on a rifle, were used in following the rocket when the speed was high. In plate 7, figure 1, which shows the clock mechanism in detail, the observer is indicating the altitude trace. This device proved satisfactory except when the trajectory of the rocket was in the plane of the tower and the telescope. For great heights, short-wave radio direction finders, for following the rocket during the descent, will be preferable to telescopes.

During this period a number of flights were made for the purpose of testing the regulation of the nitrogen gas pressure. A beginning on the problem of automatically stabilized vertical flight was also made, and the first flight with gyroscopically controlled vanes was obtained on April 19, 1932, with the same model that employed the first liquid nitrogen tank. The method of stabilization consisted in forcing vanes into the blast of the rocket[5] by means of gas pressure, this pressure being controlled by a small gyroscope.

As has been found by later tests, the vanes used in the flight of April 19, 1932, were too small to produce sufficiently rapid correction. Nevertheless, the two vanes which, by entering the rocket blast, should have moved the rocket back to the vertical position were found to be warmer than the others after the rocket landed.

This part of the development work, being for the purpose of obtaining satisfactory and reproducible performance of the rocket in the air, was conducted without any special attempt to secure great lightness, and therefore great altitudes.

In May 1932 the results that had been obtained were placed before the advisory committee, which voted to recommend the 2 additional years of the development. Owing to the economic conditions then existing, however, it was found impossible to continue the flights in New Mexico.

A grant from the Smithsonian Institution enabled the writer, who resumed full-time teaching in Clark University in the fall of 1932, to carry out tests that did not require flights, in the physics laboratories of the University during 1932-33, and a grant was received from the Daniel and Florence Guggenheim Foundation which made possible a more extended program of the same nature in 1933-34.

## RESUMPTION OF FLIGHTS IN NEW MEXICO

A grant made by the Daniel and Florence Guggenheim Foundation in August 1934, together with leave of absence for the writer granted

---

[5] U. S. Patent, Mechanism for Directing Flight, No. 1,879,187, September 27, 1932.

by the Trustees of Clark University, made it possible to continue the development on a scale permitting actual flights to be made. This was very desirable, as further laboratory work could not be carried out effectively without flights in which to test performance under practical conditions.

Work was begun in September 1934, the shop being put in running order and the equipment at the tower for the flights being re-

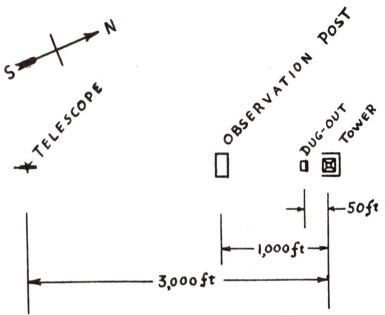

Fig. 1.—Relative positions of launching tower, dugout, shelter, and telescope.

placed. The system of remote control previously used was further improved and simplified, and a concrete dugout (pl. 7, fig. 2) was constructed 50 feet from the launching tower in order to make it possible for an observer to watch the launching of the rocket at close range. The relative positions of launching tower, dugout, shelter, and telescope are shown in figure 1.

### DEVELOPMENT OF STABILIZED FLIGHT

It was of the first importance to perfect the means of keeping the rockets in a vertical course automatically, work on which was begun in the preceding series of flights, since a rocket cannot rise vertically to a very great height without a correction being made when it deviates from the vertical course. Such correction is especially important at the time the rocket starts to rise, for a rocket of very great range must be loaded with a maximum amount of propellant and consequently must start with as small an acceleration as possible. At these small initial velocities fixed air vanes, especially those of large size, are worse than useless, as they increase the deviations due to the wind. It should be remarked that fixed air vanes should preferably be small, or dispensed with entirely, if automatic stabilization is employed, to minimize air resistance.

In order to make the construction of the rockets as rapid as possible, combustion chambers were used of the same size as those in the work of 1930-32, together with the simplest means of supplying pressure, namely, the use of a tank of compressed nitrogen gas on the rocket. The rockets were, at the same time, made as nearly streamline as possible without resorting to special means for forming the jacket or casing.

## PENDULUM STABILIZER

A pendulum stabilizer was used in the first of the new series of flights to test the directing vanes, for the reason that such a stabilizer could be more easily constructed and repaired than a gyroscope stabilizer, and would require very little adjustment. A pendulum stabilizer could correct the flight for the first few hundred feet, where the acceleration is small, but it would not be satisfactory where the acceleration is large, since the axis of the pendulum extends in a direction which is the resultant of the acceleration of the rocket and the acceleration of gravity, and is therefore inclined from the vertical as soon as the rocket ceases to move in a vertical direction. The pendulum stabilizer, as was expected, gave an indication of operating the vanes for the first few hundred feet, but not thereafter. The rocket rose about 1,000 feet, continued in a horizontal direction for a time, and finally landed 11,000 feet from the tower, traveling at a velocity of over 700 miles per hour near the end of the period of propulsion, as observed with the recording telescope.

## GYROSCOPE STABILIZER

Inasmuch as control by a small gyroscope is the best as well as the lightest means of operating the directing vanes, the action of the gyroscope being independent of the direction and acceleration of the rocket, a gyroscope having the necessary characteristics was developed, after numerous tests.

The gyroscope, shown in plate 8, figure 1, was set to apply controlling force when the axis of the rocket deviated 10° or more from the vertical. In the first flight of the present series of tests with gyroscopic control, on March 28, 1935, the rocket as viewed from the 1,000-foot shelter traveled first to the left and then to the right, thereafter describing a smooth and rather flat trajectory. This result was encouraging, as it indicated the presence of an actual stabilizing force of sufficient magnitude to turn the rocket back to a vertical course. The greatest height in this flight was 4,800 feet, the horizontal distance 13,000 feet, and the maximum speed 550 miles per hour.

In subsequent flights, with adjustments and improvements in the stabilizing arrangements, the rockets have been stabilized up to the time propulsion ceased, the trajectory being a smooth curve beyond this point. In the rockets so far used, the vanes have moved only during the period of propulsion, but with a continuation of the supply of compressed gas the vanes could evidently act against the slip stream of air as long as the rocket was in motion in air of appreciable density. The oscillations each side of the vertical varied from 10° to 30° and occupied from 1 to 2 seconds. Inasmuch as the rockets started slowly, the first few hundred feet of the flight reminded one of a fish swimming in a vertical direction. The gyroscope and directing vanes were tested carefully before each flight, by inclining and rotating the rocket while it was suspended from the 20-foot tower (pl. 8, fig. 2). The rocket is shown in the launching tower, ready for a flight. in the close-up (pl. 9, fig. 1), and also in plate 9, figure 2, which shows the entire tower.

The behavior of the rocket in stabilized flight is shown in plates 10 and 11, which are enlarged from 16-mm motion picture films of the flights. The time intervals are 1.0 second for the first 5 seconds, and 0.5 second thereafter. The 60-foot tower from which the rockets rise (pl. 9, fig. 2) appears small in the first few of each set of the motion pictures, since the camera was 1,000 feet away, at the shelter shown in plate 6, figure 1. The continually increasing speed of the rockets, with the accompanying steady roar, make the flights very impressive. In the two flights for which the moving pictures are shown, the rocket left a smoke trail and had a small, intensely white flame issuing from the nozzle, which at times nearly disappeared with no decrease in roar or propelling force. This smoke may be avoided by varying the proportion of the fluids used in the rocket, but is of advantage in following the path of the rocket. The occasional white flashes below the rocket, seen in the photographs, are explosions of gasoline vapor in the air.

Plate 10 shows the flight of October 14, 1935, in which the rocket rose 4,000 feet, and plate 11 shows the flight of May 31, 1935, in which the rocket rose 7,500 feet. The oscillations from side to side, above mentioned, are evident in the two sets of photographs. These photographs also show the slow rise of the rocket from the launching tower, but do not show the very great increase in speed that takes place a few seconds after leaving the tower, for the reason that the motion picture camera followed the rockets in flight.

A lengthwise quadrant of the rocket casing was painted red in order to show to what extent rotation about the long axis occurred in flight. Such rotation as was observed was always slow, being at the rate of 20 to 60 seconds for one rotation.

As in the flights of 1930-32 to study rocket performance in the air, no attempt was made in the flights of 1934-35 to reduce the weight of the rockets, which varied from 58 to 85 pounds. A reduction of weight would be useless before a vertical course of the rocket could be maintained automatically. The speed of 700 miles per hour, although high, was not as much as could be obtained by a light rocket, and the heights, also, were much less than could be obtained by a light rocket of the same power.

It is worth mentioning that inasmuch as the delicate directional apparatus functioned while the rockets were in flight, it should be possible to carry recording instruments on the rocket without damage or changes in adjustment.

## FURTHER DEVELOPMENT

The next step in the development of the liquid-propellant rocket is the reduction of weight to a minimum. Some progress along this line has already been made. This work, when completed, will be made the subject of a later report.

## CONCLUSION

The chief accomplishments to date are the development of a combustion chamber, or rocket motor, that is extremely light and powerful and can be used repeatedly, and of a means of stabilization that operates automatically while the rocket is in flight.

I wish to express my deep appreciation for the grants from Daniel Guggenheim, the Daniel and Florence Guggenheim Foundation, and the Carnegie Institution of Washington, which have made this work possible, and to President Atwood and the Trustees of Clark University for leave of absence. I wish also to express my indebtedness to Dr. John C. Merriam and the members of the advisory committee, especially to Col. Charles A. Lindbergh for his active interest in the work and to Dr. Charles G. Abbot, Secretary of the Smithsonian Institution, for his help in the early stages of the development and his continued interest.

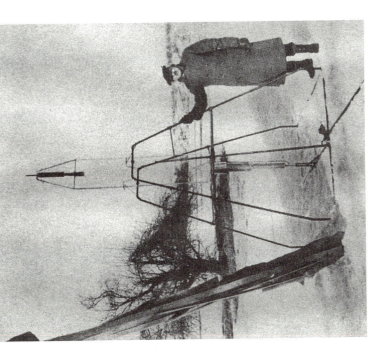

2. Assistant igniting the rocket shown in figure 1.

1. Liquid oxygen-gasoline rocket in the frame from which it was fired on March 16, 1926, in Auburn, Mass.

1. Group that witnessed the flight of the rocket shown in plate 1.

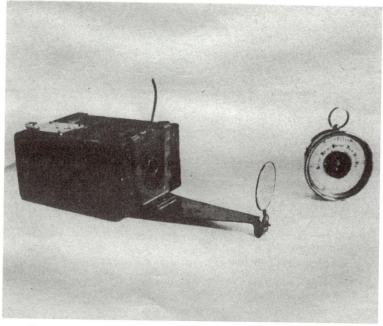

2. Barometer and camera retrieved intact after the flight of July 17, 1929.

1. Shop erected at Roswell, N. Mex., in September 1930.

2. Interior of shop.

2. 20-foot tower for static tests at Roswell, N. Mex.

1. 60-foot tower, previously used in Auburn and Fort Devens, as

1. Revolving drum to record the lift developed in static tests in the 20-foot tower.

2. Cable being unwound between the tower and the 1,000-foot shelter.

1. Control keys being operated at the shelter.

2. Observer stationed 3,000 feet from the tower with a recording telescope.

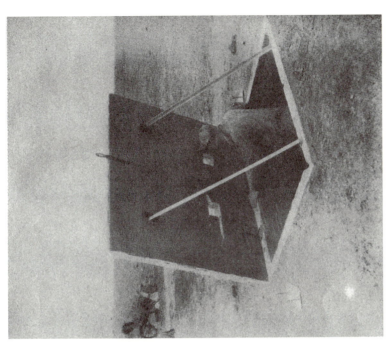

2. Concrete dugout constructed 50 feet from the launching tower so that an observer can watch the launching of the rocket at close range.

1. Clock mechanism on the recording telescope; the observer is indicating the altitude trace.

2. Testing the gyroscope and directing vanes before a flight by inclining and rotating the rocket while it was suspended from the

1. The gyroscope stabilizer.

2. Same as figure 1, except that the entire tower is shown.

1. Rocket in the launching tower, ready for a flight.

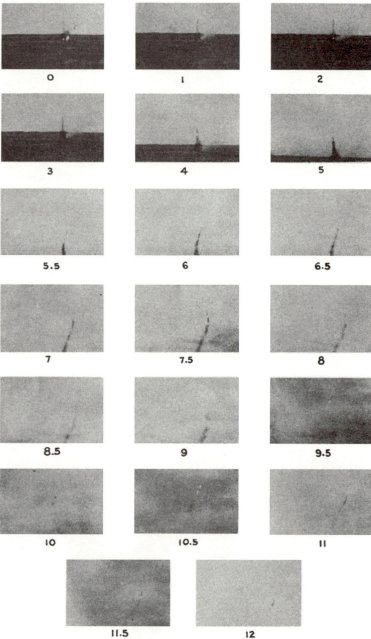

TIME IN SECONDS

The flight of October 14, 1935, in which the rocket rose 4,000 feet.

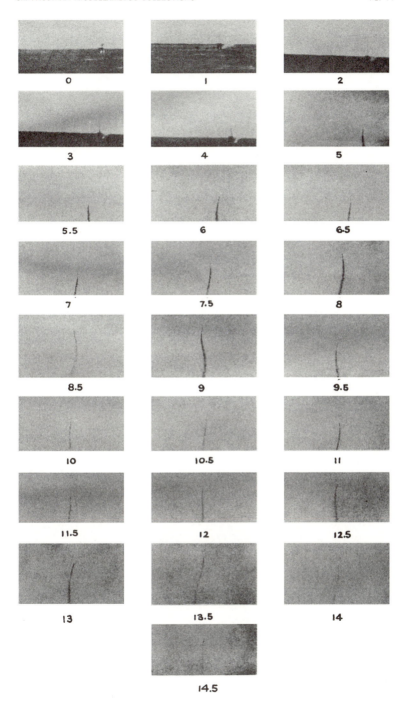

The flight of May 31, 1935, in which the rocket rose 7,500 feet.

# DR. ROBERT H. GODDARD

## A Biographical Note and Appreciation

ROBERT HUTCHINGS GODDARD was born in Worcester, Mass., on October 5, 1882. His early schooling was obtained at Boston. where he lived with his family until he was sixteen. His college work was taken at Worcester, where he was graduated from the Worcester Polytechnic Institute in 1908.

Upon graduation, he obtained a position at Worcester Polytechnic Institute as an instructor in physics. He continued to be connected with the academic world until 1943, part of the time on leaves of absence. His teaching career was conventional, rising in the usual steps from instructor to assistant professor and finally to full professor at Clark University. During a small part of this period, in the 1912-1913 season, he served as research fellow at Princeton University. The rest of his academic career was passed in Worcester.

In his school days Goddard enjoyed mathematics, and was fond of studying better ways to do things. In his freshman year at college one of his professors assigned the topic "Traveling in 1950" as a theme subject. Goddard produced a bold paper in which he described in detail a railway line between Boston and New York, in which the cars were run in an evacuated tube and were prevented from metal-to-metal contact with the guide rails by electromagnets. With such a "vacuum railroad" he calculated it would be possible to make the run from New York to Boston in ten minutes.

As a young professor of physics, Goddard made contributions of importance on the conduction of electricity from powders, the development of crystal rectifiers, the balancing of airplanes, and the production of gases by electrical discharges in vacuum tubes. During his fellowship in Princeton, he produced the first laboratory demonstration of mechanical force from a "displacement current" in a magnetic field; this current being the fundamental concept in Maxwell's theory of electromagnetic waves (radio).

These achievements, however, were merely tune-ups for the real accomplishments of his life, which were soon to begin. There is apparently no record of the first experiments he made with rockets, though it is known that this work began as part of a search for practical means of sending meteorological instruments into the stratosphere. To friends, Dr. Goddard once described with amusement some static tests he made as early as 1908 with small rockets in the basement of Worcester Polytechnic Institute. His experiments filled the basement of the building with acrid smoke, and so disturbed the equanimity of the institution that he was asked to desist, at least until better equipment could be provided.

It was during his brief period at Princeton in 1912 that he made the initial computations which later were to form the basis of the Smithsonian paper of 1919. In this period, when he was about thirty, the great excitement of discovery first began to come upon him, for his calculations clearly indicated that only a little fuel, relatively, would be required to lift a payload to really great heights by rocket power, provided the rocket were so constructed as to make use of the fuel effectively.

Upon returning to Clark, in 1914, he began to experiment in earnest, beginning with ship rockets, and continuing with rockets of various types manufactured by himself. By 1916 he had reached the limit of what he could do on his own resources. Inexperienced though he was in the ways of money-raising for scientific research, his earnestness and enthusiasm won respect and attention. When he presented his idea on paper to the Smithsonian Institution that year he promptly received a letter from Dr. Charles D. Walcott, then secretary of the Institution, commending him on the report and inquiring how much money would be needed.

Goddard guessed it would require $10,000, but cautiously asked for $5,000. Between that day in 1916 and the appearance of his first paper in 1919, the experimental work required a total of $11,000, the whole sum of which was made available by the Smithsonian. This was the investment that launched modern rocketry and jet propulsion.

The rest of Goddard's achievements are told, factually, in the two reports republished in this book. What is not disclosed—what can never adequately be told—is the labor and persistence and thought and heartbreak that went into these accomplishments, through which Goddard fathered all the research and development which led to the great expansion of jet propulsion in World War II; which continues to grow and unfold today in the jet propulsion research achievements of peace.

In 1924 Goddard married Esther Christine Kisk, who had been associated with him in his work, and who continued after their marriage to play a large part in the continuance of his research. He frequently ascribed to her the courage and faith which made his continuing efforts possible. Among other things, she was the official photographer of his tests. It was her camera that produced the pictures which illustrate his second report.

After the entry of the United States into the first World War in 1917, Goddard volunteered his services, and was set to the task of exploring the military possibilities of rockets. He succeeded in developing a trajectory rocket which fired intermittently, the charges being injected into the combustion chamber by a method similar to that of the repeating rifle. He also developed several types of projectile rockets intended to be fired at tanks or other military objectives, from a launching tube held in the hands and steadied by two short legs, a device similar in many respects to the "bazooka" of World War II.

These weapons were demonstrated at the Aberdeen Proving Grounds on November 10, 1918, before representatives of the Signal Corps, the Air Corps, the Army Ordnance and others. The demonstrations went off quite successfully, but the Armistice next day put an end to the war and also to the experiments.

In the Second World War Goddard likewise volunteered his services, and was engaged in liquid fuel rocket research for the Navy at Annapolis throughout the conflict.

Goddard concluded his last report, in 1936, with these words: "The next step in the development of the liquid-propellant rocket is the reduction of weight to a minimum. Some progress along this line has already been made."

Part of this progress consisted of the development of ingenious, light-weight, simple fuel pumps for injecting the propellants rapidly into the liquid-fuel rocket motor. The physicist had expected to return to New Mexico as soon as possible after the War, to continue his work on high altitude rockets, and planned to set some altitude records which would have been spectacular indeed.

His death, at the age of 62, brought this program to an untimely end. Nevertheless, Goddard lived to see the dream of his youth become reality. Jet propulsion, at least for the uses of war, matured in his lifetime from a fantastic notion into a billion-dollar industry. It gave promise, too, of achieving the objectives of peacetime research for which he had spent a lifetime of thought and effort.

Dr. Goddard had been a member of the American Rocket Society for many years, and a few months before his death was elected to the Society's Board of Directors. He was universally beloved and respected, and especially so by his associates in research on rockets and jet propulsion. The Board of Directors of the American Rocket Society paid tribute to him in these words:

"With the death on August 10, 1945, of Dr. Robert H. Goddard, American science has lost one of its greatest pioneers—the creator of the modern science of rocketry.

"His investigations covered almost every essential principle involved in both the theory and practice of high-power rockets, and were mainly responsible for the immense progress of the subject in the last three decades, which has exceeded in importance the results previously attained in several centuries of early development.

"His inventions included the first liquid-fuel rocket, the first smokeless powder rocket, the first practical automatic steering device for rockets, and innumerable other devices. He was one of the first to develop a general theory of rocket action, including the important "optimum velocity" principle, and to prove experimentally the efficiency of rocket propulsion in a vacuum.

"Even more impressive than Dr. Goddard's technical skill and ingenuity was his extraordinary perseverance, patience and courage in carrying on his investigations in the teeth of public skepticism and indifference, with limited financial resources, and in spite of heartbreaking technical difficulties—a combination of obstacles which might have baffled and disheartened a less stout-hearted pioneer. Almost single-handed, Dr. Goddard developed rocketry from a vague dream to one of the most significant branches of modern engineering.

"The lifework of Dr. Goddard, both as a scientist and a man, will always remain a brilliant inspiration to those who are privileged to carry on his endeavors, and to every other bold explorer on the new frontiers of science. In time to come, his name will be set among the foremost of American technical pioneers."

G. Edward Pendray

August 15, 1945